GLOBAL CLIMATE CHANGE REVISITED

Global Climate Change Revisited

Horace M. Karling
Editor

Nova Science Publishers, Inc.
New York

For permission to use material from this book please contact us:
Telephone 631-231-7269; Fax 631-231-8175
Web Site: http://www.novapublishers.com

LIBRARY OF CONGRESS CATALOGING-IN-PUBLICATION DATA
Global climate change revisited / Horace M. Karling, editor.
p. ; cm.
Includes index.
ISBN-13: 978-1-59454-039-4 (hardcover)
ISBN-10: 1-59454-039-X (hardcover)
1. Climatic changes. I. Karling, Horace M.
QC981.8.C5G6525 2006
363.738'74—dc22
2007008719

Published by Nova Science Publishers, Inc. ✦ New York

Contents

In: Global Climate Change Revisited
Editor: Horace M. Karling, pp. 1-25

ISBN 978-1-59454-039-4

Chapter 1

CLIMATE CHANGE: DESIGN APPROACHES FOR A GREENHOUSE GAS REDUCTION PROGRAM*

Larry Parker

SUMMARY

With the passage of the 2005 Sense of the Senate climate change resolution calling on the Congress to enact a mandatory, market-based program to slow, stop, and reverse the growth of greenhouse gases, the issue of related costs has taken on increased importance. Indeed, the resolution itself states that the program should be enacted at a rate and in a manner that "will not significantly harm the United States economy" and "will encourage comparable action" by other nations. Facets of the cost issue that have raised concern include absolute costs to the economy, distribution of costs across industries, competitive impact domestically and internationally, incentives for new technology, and uncertainty about possible costs.

In general, market-based mechanisms to reduce greenhouse gas emissions, the most important being carbon dioxide (CO_2), focus on specifying either the acceptable emissions level (quantity) or compliance costs (price), and allowing the marketplace to determine the economically efficient solution for the other variable. For example, a tradeable permit program sets the amount of emissions allowable under the program (i.e., the number of permits available limits or caps allowable emissions), while allowing the marketplace to determine what each permit will be worth. Likewise, a carbon tax sets the maximum unit cost (per ton of CO_2 equivalent) that one should pay for reducing emissions, while the marketplace determines how much actually gets reduced. In one sense, preference for a carbon tax or a tradeable permit system depends on how one views the uncertainty of costs involved and benefits to be received.

Market-based mechanisms attempt to address the cost issue by introducing flexibility into the implementation process. The cornerstone of that flexibility is permitting sources to decide for themselves their appropriate implementation strategy within the parameters of market

* Excerpted from CRS Report RL 33799 dated January 16, 2007.

signals and other incentives. That signal can be as simple as a carbon tax or comprehensive credit auction that tells the emitter the value of any reduction in greenhouse gases, to a credit marketplace that is constrained by a ceiling price (safety valve) and includes incentives for new technology. As illustrated here, the combinations of market mechanisms are numerous, allowing decision makers to tailor the program to address specific concerns.

In a sense, the options discussed here represent a continuum between alternatives focused on the price side of the equation (e.g., carbon taxes) through hybrid schemes (e.g., safety valves) to alternatives focused on the quantity side (e.g., banking and borrowing). They are tools to assist in the assessment of potential greenhouse gas reduction approaches, leaving any policy decision on balancing the price-quantity issue to the ultimate decision makers.

Climate change has been a continuing policy issue since the United States ratified the 1992 United Nations Framework Convention on Climate Change (UNFCCC). An integral part, and sometimes driving part, of the ensuing debate has been the issue of cost — in several manifestations.[1] For the George W. Bush Administration, the Kyoto Protocol was "fatally flawed in fundamental ways," including requiring compliance with mandates that "would have a negative economic impact with layoffs of workers and price increases for consumers."[2] This concern about cost can also be seen in the Senate's most recent resolution on climate change in 2005 (S.Amdt. 866). Echoing the language of its 1997 resolution (S.Res. 98) on the same subject, the 2005 Sense of the Senate resolution on climate change declared that a mandatory, market-based program to slow, stop, and reverse the growth of greenhouse gases[3] should be enacted at a rate and in a manner that "will not significantly harm the United States economy" and "will encourage comparable action" by other nations.[4] Facets of the cost issue that have raised concern include absolute costs to the economy, distribution of costs across industries, competitive impact domestically and internationally, incentives for new technology, and uncertainty about possible costs.

Because a stalemate has persisted on strategies to control greenhouse gas (GHG) emissions, particularly because of cost uncertainties, attention has increasingly focused on options to address these concerns and to move the debate forward. These options range from incremental mechanisms within a tradeable permit program, such as banking and borrowing of credits, that have a minimal effect on overall emission reduction targets goals to more fundamental proposals, such as a carbon tax, that would take climate change policy in a new and somewhat uncharted direction.

This paper explores these options to address the cost issue in four parts. First, the basic economic tradeoff between controlling the quantity of GHG emissions and the program's compliance costs is introduced and explained. Second, the five dimensions of the cost issue that have arisen so far in the climate change debate are identified and discussed. Third, a representative sample of proposed aproaches to address cost concerns is compared and analyzed according to the five cost dimensions identified previously. Finally, the proposed options are summarized and opportunities to combine or merge different approaches are analyzed. The paper does not provide a detailed discussion of allocation and implementation issues that creating a market-based mechanism (particularly a cap-and-trade program) would entail.

INTRODUCTION: THE PRICE VERSUS QUANTITY DEBATE

In general, market-based mechanisms to reduce GHG emissions, the most important being carbon dioxide (CO_2), focus on specifying either the acceptable emissions level (quantity) or compliance costs (price) and allowing the marketplace to determine the economically efficient solution for the other variable. For example, a tradeable permit program sets the amount of emissions allowable under the program (i.e., the number of permits available limits or caps allowable emissions), while permitting the marketplace to determine what each permit will be worth. Likewise, a carbon tax sets the maximum unit cost (per ton of CO_2 equivalent) that one should pay for reducing emissions, while the marketplace determines how much actually gets reduced. In one sense, preference for a carbon tax or a tradeable permit system depends on how one views the uncertainty of costs involved and benefits to be received.

For those confident that achieving a specific level of CO_2 reduction will yield significant benefits — enough so that even the potentially very high end of the marginal cost curve does not bother them — a tradeable permit program may be most appropriate. CO_2 emissions would be reduced to a specific level, and in the case of a tradeable permit program, the cost involved would be handled efficiently, though not controlled at a specific cost level. This efficiency occurs because through the trading of permits, emission reduction efforts focus on sources at which controls can be achieved at least cost.

However, if one feels more certain of the potential downside risk of substantial control costs to the economy than of the benefits of a specific level of reduction, then a carbon tax may be most appropriate. In this approach, the level of the tax effectively caps the marginal cost of control that affected activities would pay under the reductions scheme, but the precise level of CO_2 reduction achieved is less certain.

Emitters of CO_2 would spend money controlling CO_2 emissions up to the level of the tax. However, because the marginal cost of control among millions of emitters is not well known, the overall emissions reductions for a given tax level on CO_2 emissions is subject to some uncertainty.

Hence, a major policy question is whether one is more concerned about the possible economic cost of the program and therefore willing to accept some uncertainty about the amount of reduction received (i.e., carbon taxes); or one is more concerned about achieving a specific emission reduction level with costs handled efficiently, but not capped (i.e., tradeable permits).

A model for a tradeable permit approach is the sulfur dioxide (SO_2) allowance program contained in title IV of the 1990 Clean Air Act Amendments (42 U.S.C. 7651). Also called the Acid Rain Program, the tradeable permit system is based on two premises. First, a set amount of SO_2 emitted by human activities can be assimilated by the ecological system without undue harm. Thus the goal of the program is to put a ceiling, or cap, on the total emissions of SO_2 rather than limit ambient concentrations. Second, a market in pollution licenses between polluters is the most cost-effective means of achieving a given reduction. This market in pollution licenses (or allowances, each of which is equal to 1 ton of SO_2) is designed so that owners of allowances can trade those allowances with other emitters who need them or retain (bank) them for future use or sale. Initially, most allowances were allocated free by the federal government to utilities according to statutory formulas related to a given facility's historic fuel use and emissions; other allowances have been reserved by the government for periodic auctions to ensure market liquidity.

There are no existing U.S. models of an emissions tax, although five European countries have carbon-based taxes.[5] The closest U.S. example is the tax on ozone-depleting chemicals (ODCs). To facilitate the phaseout of ODCs under the Montreal Protocol and subsequent amendments, the United States imposed a tax on specific ODCs in 1990. This tax was designed to supplement the allowance trading program that the EPA had designed to implement the international agreements. Several activities trigger the tax, including the production and import of the chemicals, along with the importation of products that contained them or used them in their production processes. In addition, inventories of certain ODCs held on January 1 of each year are subjected to a "floor stocks tax."[6]

FIVE DIMENSIONS OF THE COST ISSUE

Five dimensions of costs associated with reducing GHG emissions are discussed in this section: (1) absolute costs, (2) distribution of costs, (3) long-term costs, (4) price signal and stability, and (5) uncertainty of costs.

The *absolute costs* of a GHG reduction program are a function of the interplay among the tonnage reduction required, the timetable imposed on that reduction, and the techniques available and used to achieve that reduction (the "three Ts"). Variables involved with the tonnage requirement include the magnitude and firmness of the reduction requirement and the number of gases and sectors involved in the program. Variables involved with the timetable include its length and number of phases, along with the number and extent of any deadline extensions allowed and on what basis. Finally, variables involved with techniques include promotion and availability of new technology, the degree of flexibility permitted in complying with the program, and any ceiling on compliance costs. All these program design parameters influence the absolute cost of the program and the timing and extent of any benefits received.

A second concern with costs is their *distribution* across the various sectors of the economy. As indicated by Table 1, GHG emissions are spread throughout the economy, with about 81% emitted by the electric power, transportation, and industry sectors. Restricting participation by any group could increase the absolute cost of the program and would certainly increase the costs to the remaining participants. However, numerous rationales have been put forward to justify excluding one group or sector from a reduction requirement, or to provide some other special consideration. Rationales offered include a sector or industry's concern about (1) international competitiveness, (2) lack of cost-effective control options, (3) inability to make necessary capital investments, (4) economic disruption, (5) credit for previous efforts that reduced emissions, and (6) the "minor" contribution that industry or sector makes to the overall problem. It is the multitude of such variables that make constructing an acceptable reduction allocation scheme very difficult.

Table 1. 2004 U.S. Greenhouse Gas Emissions

Economic Sector	Million Metric Tons of CO_2 equivalent	Percentage of Total
Electric power industry	2,338	33.3%
Transportation	1,955	27.9%
Industry	1,377	19.6%
Agriculture	491	7.0%
Commercial	460	6.6%
Residential	391	5.6%
Total	7,012	100.0%

Source: U.S. Environmental Protection Agency, *Inventory of U.S. Greenhouse Gas Emissions and Sinks: 1990-2004*, EPA 430-R-06-002 (Washington, DC), p. ES-13.

Note: The total does not include 62 million metric tons from U.S. territories.

A third concern is the *long-term cost* considerations of a GHG reduction program. Climate change policy has to be thought of in decades, not years. Ultimately, a successful climate change program would involve a long-term transition to a less carbon-emitting economy. Generally, studies that indicate the availability and cost-effectiveness of emerging new technologies to achieve this transition include an economic mechanism to provide the necessary long-term price signal to direct research, development, demonstration, and deployment efforts.[7] Developing such a price signal involves variables such as the magnitude and nature of the market signal, and the timing, direction, and duration of it. In addition, studies indicate combining a sustained price signal with public support for research and development efforts is the most effective long-term strategy for encouraging development of new technology.[8] As stated by Morgenstern: "The key to a long term research and development strategy is both a rising carbon price, and some form of government supported research program to compensate for market imperfections."[9]

A fourth consideration is the stability of the *price signal* in whatever form it takes (e.g., allowance prices, carbon taxes, auction prices). A stable and reliable signal is necessary to minimize economic disruption and to encourage new technology. Experience with existing emissions markets suggests that short-term price spikes and troughs occur that have at least short-term economic effects, either disrupting the market (in the case of high prices) or discouraging new technology (in the case of low prices). Causes of this volatility can include (1) lack of trading volume, (2) illiquidity in the market, (3) external events, and (4) regulatory uncertainty. History with previous emissions trading programs suggests that if a greenhouse gas program is based on a market-based implementation strategy, the inclusion of flexibility mechanisms to ensure reasonable market stability is desirable.[10]

A final cost consideration is the *cost uncertainty* presented by the wide range of projected costs of GHG reduction. To the extent one understands the variables that create the range presented by different forecasting models, one can design a program to address those variables. Projected costs of a proposed greenhouse gas reduction program will differ among models, based on the various economic and technological assumptions either embedded in the particular model's processes (endogenous variables) or assumed externally and inserted into the model. Weyant has identified five assumptions that explain many of the differences in greenhouse gas reduction program cost estimates:[11]

- Basecase projections of future GHG emissions and climate damages.
- The specifics of the reduction program examined (particularly the amount of flexibility permitted in complying with its mandates).
- How dynamic the model is, representing substitution possibilities by producers and consumers, including the turnover of capital equipment.
- How the rate and processes of technological change are modeled.
- How benefits are modeled.

Figure 1, below, illustrates how these and other variables (such as type of model used) can influence the estimated costs of climate change legislation. Measured by impact on GDP, the figure indicates impacts ranging from a positive 2% increase in GDP to a 4% decrease. Interestingly, the variables used in projecting cost and benefits are sufficiently robust to obscure a strong correlation between cost and reduction requirements.

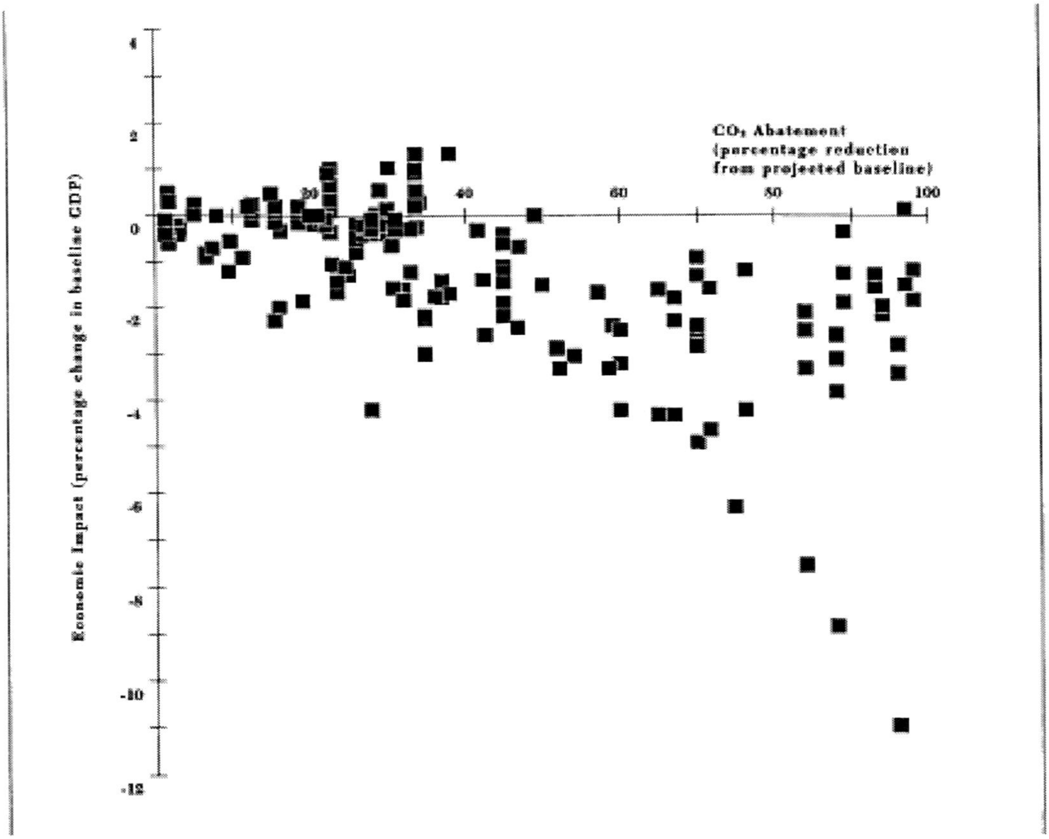

Source: Robert Repetto and Duncan Austin, *The Costs of Climate Protection: A Guide for the Perplexed* (World Resources Institute, 1997), p. 12.

Figure 1. The Predicted impacts of carbon abatement on the U.S. economy (162 estimates from 16 models.

The range indicated also reflects the perspectives and parameters assumed by the forecast authors. In a previous report, CRS noted that cost analyses are influenced by the perspective (or lens) through which one views the problem.[12] Analysts viewing climate change policy through a technological perspective see it as an impetus for improved efficiency through technology improvements in the economy, consistent with concepts such as life-cycle costs. Analysts viewing policy through an economic lens work through the boundaries of market economics and cost-benefit considerations. Finally, analysts viewing the issue through an ecological lens look to the benefits of controlling greenhouse gases and are suspicious of "baseline" scenarios that suggest that "business as usual" is an acceptable yardstick from which to measure policy changes. Each of these lenses implies fundamentally different ways of assessing policy actions and modeling potential costs and benefits. The quantitative result are cost estimates that range from actual savings to the economy (from GHG reductions) to substantial costs.

APPROACHES FOR ADDRESSING COST CONCERNS

The following analysis of options to address the cost concerns identified above is loosely arranged by the focus of the specific option: (1) the tonnage requirement, (2) the time frame, and (3) the techniques allowed for compliance. It should be noted that several options examined affect more than one of the "Ts." Also, the options are not mutually exclusive — many can be combined to create more refined options.

Tonnage Options

Much of the discussion on GHG reductions has focused on a historic baseline as the starting point for reductions. Assuming that the emissions inventory for a specific year is adequate to support a regulatory program (whether market-based or not), such a baseline is reasonable. Most existing emissions trading programs are based on a historic baseline with modifications. However, there are options to calculate a baseline that responds to economic events over time without necessarily compromising the tonnage cap. Also, the historic baseline can be eliminated in favor of different methods of achieving specific reductions.

Carbon Tax

The most radical approach to controlling costs and addressing the concerns identified above is to impose a carbon tax in lieu of proposed allowance trading programs. As discussed in the "Introduction," under a carbon tax, the costs are fixed by the legislation and the quantity of emissions reduced becomes the variable. Carbon taxes are generally conceived as a levy on natural gas, petroleum, and coal, according to their carbon content, in the approximate ratio of 0.6 to 0.8 to 1.0, respectively. However, the levy would not have to be imposed on the fuels themselves — proposals have been made to impose the tax downstream at the point where the fuel is converted into heat and CO_2. In addition, there is no reason why the tax could not be expanded to include all greenhouse gases in appropriate carbon equivalents.

A carefully designed carbon tax could potentially address all five of the concerns identified above. A carbon tax puts a limit on absolute cost by capping the marginal costs that participants should pay to reduce GHG emissions. Participants would receive a firm price signal with respect to the upper value of GHG emissions, and respond in the most cost-effective manner — that is, reduce emissions up to the cost of the carbon tax and pay the tax on any remaining emissions that are more expensive to eliminate.

A carbon tax can be tailored to address distributional concern in two ways. The first would be to exempt, either partly or completely, whatever sectors or industries were felt to be threatened, either competitively or otherwise, by imposing the tax. The current tax code provides numerous exemptions from various taxes for a variety of reasons. However, such an approach would create economic distortions and complicate the tax structure. The second approach would be to use some of the revenue generated by the tax to provide appropriate relief to targeted sectors or industries. This could involve increasing funding for existing programs for such sectors or industries, or creating new ones. In some ways, this approach might be more transparent than an approach that involves a potentially complicated tax structure. These approaches are not mutually exclusive; they could be combined if considered appropriate.

Likewise, a carbon tax can be employed to address long-term concerns in two ways. First, the carbon tax would create a long-term price signal to stimulate innovation and development of new technology. This price signal could be strengthened if the carbon tax were escalated over the long run — either by a statutorily determined percentage or by an index (such as the producer price index). Second, some of the revenue generated by the tax could be used to fund research, development, demonstration, and deployment of new technology to encourage the long-term transition to a less-carbon-intensive economy.

A carbon tax's basic approach to controlling GHG emissions is to supply the marketplace with a stable, consistent price signal — the fourth cost concern. Designed appropriately, there would be little danger of the price spikes or market volatility that can occur in the early stages of a tradeable permit program.

Finally, a carbon tax basically places an upper boundary on projected economic cost uncertainty. However, it increases uncertainty with respect to environmental benefits by making emission reductions a dependent variable. This is the basic tradeoff that a price-based control system presents. One way that might mitigate the problem to some extent would be to combine the carbon tax with some form of quantity controls. As noted earlier, the CFC program attached a tax to its trading program with beneficial results. However, it is the trading program, not the CFC tax, that is the primary regime for control. In this manner, a carbon tax would be more of a revenue raiser than a control regime. A second hybrid would be a "safety valve" that capped allowance prices such as proposed by the National Commission on Energy Policy.[13] That approach is discussed later in this report. The degree to which the problem is mitigated (and others created) depends on the interplay between the quantity control and the carbon tax.

Dynamic Tonnage Target

Another approach to address some of the concerns identified above is to calculate the tonnage target based on economic or other indexes or measures rather than strictly on a historic or other static baseline. For example, the National Commission on Energy Policy recommended that the tonnage requirements for a GHG reduction program begin with year 2000 emissions, with the future trajectory of emissions based on the product of a progressively declining limit on the country's GHG intensity times projected economic growth. Over time, the progressively more stringent carbon intensity index would produce progressively more stringent emission tonnage caps, despite projected increases in economic growth. The actual steepness of that path would depend on the rate of decline in carbon intensity mandated by the program and actual economic growth. Of course, the dynamic tonnage target could be indexed to just about any relative variable (e.g., energy prices).

Depending on the specifics of the methodology and measures used in creating it, the dynamic target could be more responsive to some unforeseen events, such as substantial economic growth, than a static baseline. At least in the short term, this could reduce costs and economic disruption if a sharp spike in economic growth were to occur. In contrast, slower-than-anticipated growth would reduce the available emission credits and thereby reduce the potential for "hot air" credits (i.e., credits "created" by a slowdown in the economy rather than by control efforts).[14]

By potentially mitigating some effects of a static, historic emissions baseline, a dynamic tonnage methodology allows flexibility in distributing reductions and the resulting costs among different sectors of the economy. Growth, GHG intensity, production, and other variables could be tailored for sectors, states, or regions based on specific concerns, such as competitiveness. For example, an industry growth index could be used to calculate reduction requirements rather than an aggregated index such as GDP. Like most schemes, a dynamic target scheme could completely exclude some industries, with the obvious result of a shift in cost to the ones remaining in the program.

The effect of a dynamic target on long-term costs would depend on the slope of reductions mandated by the program. For example, the recommendation of the National Commission on Energy Policy called for an annual 2.4% reduction in allowable GHG intensity increasing to 2.8% annually after 10 years. This declining curve would be multiplied by a projection of presumably increasing economic growth. A steeper slope in GHG intensity mandates and/or an overly pessimistic projection of economic growth would strengthen the need for less carbon-intensive technology, but at the risk of increasing cost if those technologies did not arrive in a timely manner. A weak GHG intensity mandate and/or an overly optimistic projection of economic growth could reduce necessary emission reductions and provide a weak incentive for new technology.

A dynamic target would not necessarily prevent short-term fluctuations in the price signal, depending on the frequency of adjustments. If a target was based on macro-economic trends, such as GDP, it would not respond much to short-term or localized events, such as the 2000-2001 electricity shortage in California. Also, the mixture of indices with different vectors (e.g., GHG intensity reducing targets while economic growth is increasing targets) may create some uncertainty in markets regarding the appropriate price of credits.

Finally, a dynamic target would not increase the certainty of cost estimates. Uncertainty about the future trajectory of economic growth would be reflected in cost estimates (just as they are now, with emissions capped at historically determined levels). Likewise, benefit certainty would not improve for the same reason.

Expand Supply Options

The breadth of options permitted under a reduction program can have a significant effect on absolute costs. Legislation introduced in recent Congresses has ranged from programs based on one economic sector (e.g., electric utilities) and one greenhouse gas (e.g., carbon dioxide) to several sectors (including opt-in provisions) and all six greenhouse gases covered by the Kyoto Protocol.[15] Also, some proposed programs have included international trading of emission credits and biological sequestration offsets among the permissible means of complying with reduction requirements.[16] Some of these options, particularly international trading and sequestration, have included limits on their applicability. For example, the Regional Greenhouse Gas Initiative[17] has put control cost triggers (characterized as "safety valves") on the availability of some supply options, such as sequestration.

Numerous analyses were done on the impact of global trading after the signing of the 1997 Kyoto Protocol. For the United States, the cost of complying with the Kyoto Protocol was estimated at $23-$50 per ton of carbon if global trading were included, versus $61-$119 if only trading among developed (Annex 1) countries were permitted. Cost estimates of "No trading" scenarios ranged from $193-$295 per ton.[18] Studies have suggested that, beyond international trading, including non-carbon dioxide greenhouse gases and sequestration in the supply mix can play an important and cost-effective role in any climate change program.[19]

Expanding supply sources could help industries that do not have readily accessible means of reducing greenhouse gases on their own by providing them with additional options and making the credit market more liquid. To the extent the expanded supply sources help create an integrated market with a true market price for credits, industries could avoid very high compliance costs and lessen the impact of those costs on their profitability. However, if competitors in other countries do not have to reduce emissions at all (as is currently the case with the Kyoto Protocol), competitive disadvantage would remain in some cases.

The degree to which expanded supply options would contribute to a long-term and stable price signal would depend on how integrated these sources are in the overall permit market. For example, with the European Trading System (ETS), there are separate markets for credits created within the 15 members of the European Union (EU) covered by the EU bubble, credits created by Joint Implementation with eastern European countries, and credits created via the Clean Development Mechanism (CDM) with Third World countries.[20] One type of credit can not be traded for another. The result is a range of credit prices, reflecting the relative risk and availability of the various credit types. Thus, the long-term signal being delivered is currently unclear, and may take time to develop. Likewise, substantial fluctuations in the EU credit market have not been stabilized by the existence of the other two credit types.

In some ways, expanding supply options may increase the uncertainty of cost estimates, not only because of disparity in assumed reduction costs, but also in assumed availability and penetration of the options themselves. For example, emission reductions via the Clean Development Mechanism could be substantial and very cost-effective. However, the mechanism itself creates uncertainty with respect to availability, as does the willingness of foreign governments to participate. It is difficult to quantify the effect such an option could have on costs without some track record, as is slowly being built by the ETS.

Timetable Options

Similar to the country's three-plus decades effort to reduce smog, climate change promises to be an effort measured in decades, not years. Unlike conventional pollution control efforts, the environmental benefit of mitigating climate change would come from a reduction in the stock of greenhouse gases that have built up in the atmosphere for decades, whereas the economic costs of control are related to the current flow of additional gases into the atmosphere. Thus, in a situation similar to protecting the stratospheric ozone layer, there would be a substantial delay between control costs and environmental benefits. Indeed, if short-term reductions in the stock of greenhouse gases were the focus of climate change policy, control efforts would be focused on controlling methane, which has a 20-year lifetime, compared with CO_2, which has a 200-year lifetime. Likewise, temporary measures, such as biologic sequestration, would be accelerated with the assumption that new technology would be available in the future to capture the biologically sequestered carbon dioxide when it is released decades from now.

This situation leads to disputes over how time should be managed under a GHG reduction program. One argument is that modest cuts (or slowing of the increase) early, followed by steeper cuts later, is the most cost-effective. Generally, three cost-related arguments are made in favor of this approach. First, over the long-term, sustained GHG reductions involve a turnover in existing durable capital stock — a costly process. If the time frame of the reduction is long enough to permit that capital stock to be replaced as it wears out, the transitional costs are reduced. Second, increased time to comply would permit the development and deployment of new, less carbon-intensive technologies that are more cost-effective than existing technology. Third, assuming a positive rate of return on current investment, less money needs to be set aside today to meet those future compliance costs.[21]

A counter argument to the above focuses on the risks of delay, both in terms of scientific uncertainty and technology development. In terms of scientific uncertainty, there is no consensus on what concentration of greenhouse gases should not be exceeded in order to avoid undesirable climate change. If the stabilization level needed is relatively low, any delay in beginning reductions could be costly, both economically and environmentally.[22] Secondly, given the sometimes long lead times for technology development, both a long-term price signal and research and development funding may have to be initiated quickly to encourage technology development and deployment in time to hold GHG concentrations to a level that avoids unacceptable damages. In the same vein, an early signal with respect to climate change policy is necessary to discourage investment in durable long-lived (50-60 years) carbon-intensive technologies.[23] As stated by Jaccard and Montgomery:

> The window of opportunity for reducing cost implies a need for immediate and continuing action to develop new low-carbon technologies and to begin shifting long-lived investment decisions toward alternatives that lower carbon emissions. Absent these actions, the rapid future emissions reductions included in the delayed emissions scenario may be more costly than more evenly paced, and earlier reductions.[24]

Economic-Based Circuit Breaker

Delaying or suspending compliance with environmental mandates because of energy and economic reasons is not a novel idea. The Clean Air Act contains provisions permitting the President, in response to a petition by an affected state's governor, to temporarily suspend any part of a state implementation plan or enforcement of the SO2 trading program, to address a severe national or regional energy emergency.[25] For example, during the 2000-2001 California energy crisis, President Clinton directed all federal agencies to do their part to assist the state in meeting its electricity demand. For its part, the EPA revised its guidance on emergency generators to allow backup generators to be used to avert a power blackout.[26]

Previously, backup generators could be used only when the power was actually interrupted. The increased flexibility permitted by the EPA during the emergency meant more power at the expense of more pollution (particularly of carbon monoxide and nitrogen oxides).

Likewise, market-based systems are not immune to being suspended if economic or energy conditions turn severe. A contributing factor in the California power crisis was the Regional Clean Air Incentives Market (RECLAIM), a credit trading system for reducing nitrogen oxide (NOx) emissions. RECLAIM was established in 1994 to provide flexibility for companies in the South Coast (Los Angeles) area as controls on NOx, a major contributor to smog formation, were tightened. Because of record electricity demand in 2000, electric generators in the South Coast area generated more power than they did in the base period, resulting in utilities buying RECLAIM trading credits in unprecedented quantities. As a result, the price of credits rose from less than $1 per pound of NOx in January 2000 to more than $60 per pound of NOx by March, 2001.[27] To solve this problem, in March 2001, the South Coast Air Quality Management District amended RECLAIM to remove large power plants from the trading system and required owners of such facilities to reduce emissions under a mandated command-and-control regime. Such facilities are scheduled to return to the trading system in 2007.[28]

Proposals have been made to formalize a "circuit breaker" into any GHG reduction program. In general, proposals envision a declining emission cap system where the rate of decline over time is determined by the market price of permits. If permit prices remain under set threshold prices, the next reduction in the emissions cap is implemented. If not, the cap is held at the current level until prices decline.[29] Such a cap could be implemented on an economy-wide basis or by sector or other relevant grouping.

Because the conditional reduction approach attempts to turn both the price and the quantity of reductions into variables solved by the trading market, its effect on cost depends on a host of variables — most obviously the profiles of the emission reduction targets and the price triggers. For example, the price trigger could be based on the spot-market price, the long-term market price, or some hybrid price mechanism. Also, the reliance on the market to either directly or indirectly determine price and quantity puts pressure on regulators to oversee operations and prevent any market manipulation designed to slow emission reductions.

A conditional tonnage target could address distributional issues if its tonnage targets and timetable triggers are tailored for specific sectors or industries. This would substantially increase the complexity of the scheme and potentially risk bifurcation of the permit market. As with other permit schemes discussed here, another approach to addressing distributional concerns under a conditional tonnage target would be to simply exempt certain sectors from its mandates.

When the control regime responds to relatively short-term events, it may not provide the long-term price signal necessary to promote long-term solutions. The elastic time frame also gives ambiguous signals for planning the appropriate pace and scope of research and development efforts. In contrast, the regime's focus on short-term economic disruption may help in damping short-term volatility in the allowance market. As noted, the responsiveness of the price and timetable triggers would determine how effective the program would be in avoiding such disruption.

In reducing the uncertainty for cost estimates, the scheme introduces about the same number of new uncertainties as it does in reducing others. The circuit breaker prevents ever increasing costs, although with some undetermined lag time. However, it is difficult to estimate absolute costs because one does not know how often it will be used.

The scheme also increases uncertainty about the future trend in benefits by making the quantity of emissions reduced a variable. A short-term break might not make much of a difference, particularly if participants were required to make up the emissions later. However, it introduces uncertainty into the system that would be difficult to quantify.

Technology-Based Timetable

Another approach to increase flexibility in the system and encourage long-term technology development would be to provide special compliance schedules for entities deploying innovative, less carbon-intensive technologies. An example of this possibility is Section 409 of Title IV of the 1990 Clean Air Act Amendments.[30] Under Section 409, utilities choosing to meet their sulfur dioxide reduction requirements by installing a qualifying clean coal technology receive a four-year extension on the program compliance deadline. During the extension, the affected emitter is allowed to operate under existing regulations and operating conditions. If the technology fails to operate as designed, the affected unit may be retrofitted with another qualifying technology or with an existing control technology. For a GHG reduction program, a qualifying technology could include geologic sequestration, emerging energy efficient technology, or advanced solar power.

A technology extension could reduce costs in two ways. First, the delay in compliance itself would reduce cost by allowing the affected company more time to gather resources and optimize a compliance plan. Second, to the extent the delay encourages more cost-effective approaches to GHG reductions, compliance cost and long-term cost would be reduced. Of course, the risk is that the delay will not result in successful technology development. Indeed, it is likely that at least some of the projects would fail — that is the nature of innovation. However, because technology development is crucial to long-term reductions in greenhouse gases, some may feel the risk is worth it.

Assistance with distributional costs under this option would depend on the opportunities for new technology in given sectors of the economy. Although some industries may have potentially cost-effective technology-fixes, such as geologic sequestration, others may involve long-term structural changes.

Of course, the focus of a technology-based timetable is to provide a long-term signal to the market encouraging new technology. Such a signal could be strengthened significantly with increased government funding of projects.

Because this option is focused on new technology, it would seem likely to have little effect on short-term price volatility. However, there may be a risk that the temporary removal of significant emitters from the market-system in response to the incentive could increase short-term volatility and uncertainty by diminishing permit demand and trading volume.

It is difficult to determine the effects of this option on cost estimates. It would depend on how widespread the assumed participation rate is.

Technique Options

Most current GHG reduction proposals assume a market-based implementation strategy — generally a permit trading program. This is not surprising, as flexibility and new technologies are considered the keys to a cost-effective implementation strategy over the long run. Generally, technique options range from making a tradeable permit program more flexible through mechanisms like banking, to creating a hybrid program where the regime shifts from a quantity-based permit program to a carbon tax, depending on defined circumstances.

Banking and Borrowing

Most existing trading programs include provisions for banking credits for either future use or future sale. Indeed, the absence of effective banking in the RECLAIM program (discussed earlier) is credited with contributing to RECLAIM's suspension during the California energy crisis. As summarized by Resources for the Future (RFF):

> Allowance banking has been an essential component of the SO_2 program. Its absence is a costly feature of the NOx programs, eroding the opportunity for cost savings from interannual trading and contributing directly to the suspension of trading in RECLAIM.[31]

Banking and borrowing reduces the absolute cost of compliance by making annual emission caps flexible over time. The limited ability to shift the reduction requirement across time allows affected entities to better accommodate corporate planning for capital turnover and technological progress, to control equipment construction schedules, and to respond to transient events such as weather and economic shocks. Generally, banking and borrowing would not have any direct impact on distributional concerns, which are more directly determined by initial allocation decisions. Banking and borrowing can help provide a long-term market signal by supporting credit prices when costs are lower than expected.[32]

The flexibility provided by banking and borrowing, as noted, can help damp short-term volatility. The degree that they help is disputed. As discussed later, some argue that banking and borrowing may provide sufficient flexibility in some cases to keep market disruptions to a minimum.[33] However, others argue that if a program involves more than modest reductions, a more robust "safety valve" is preferable.[34]

In estimating costs, banking and borrowing help smooth out the reduction requirement, as witnessed by the current acid rain program. This economically desirable effect does not necessarily reduce the uncertainty in cost estimates because estimators will make different assumptions about the extent to which banking and borrowing are used by emitters. The smoothing effect, however, has no effect on the reduction requirement (in contrast with several of the other alternatives discussed here). This is a major reason why this alternative is generally favored by those whose priority is to achieve specific reductions.

Auctioning Permits

Auctions can be used in market-based pollution control schemes in several different ways. For example, Title IV of the 1990 Clean Air Act Amendments uses an annual auction to ensure the liquidity of the credit trading program. For this purpose, a small percentage of the credits permitted under the program are auctioned annually, with the proceeds returned to the entities that would have otherwise received them. Private parties are also allowed to participate. A second possibility is to use an auction to raise revenues for a related (or unrelated) program. For example, the Regional Greenhouse Gas Initiative (RGGI) is exploring an auction to implement its public benefit program to assist consumers or pursue strategic energy purposes.[35] A third possibility is to use auctions as a means of allocating some, or all, of the credits mandated under a GHG control program. In examining a modified auction program, RFF found that an auction scheme is "dramatically more cost-effective" in allocating credits than either a grandfathered allocation method[36] or a generation performance standard[37] (GPS) approach.[38] Obviously, the impact that an auction would have on the cost dimensions identified earlier would depend on how extensively it was used in any GHG control program, and to what purpose the revenues were expended.

The cost-effectiveness of an auctioning system results from allowing the marketplace to allocate credits. However, unlike a carbon tax, the market-clearing price for credits is not limited (unless the system is combined with a safety valve, as discussed below). Hence, an auction for credits would be more expensive for specific industries than under a historically based grandfathered system, where they would receive their credits free. Likewise, the price consumers pay may be greater, depending on the companies' ability to pass on their additional costs to them. However, when the substantial revenues received by the auctions are considered, auctions are more cost-effective than grandfathered or GPS systems. As stated by RFF:

> The bottom line is that the AU [auction] approach weighs in at substantially less economic cost to society than either of the two gratis approaches to allocating allowances... . AU also provides policymakers with flexibility, through the collection of revenues that can be used to meet distributional goals or to enhance the efficiency of the AU even further by reducing pre-existing taxes. Because the AU approach is so cost-effective, a corresponding a [sic] carbon policy will have less effect on economic growth than under the other approaches. This attribute provides the most significant form of distributional benefit.[39]

As noted by RFF, the revenues from an auction can be used to address a host of distributional concerns. Indeed, as noted earlier, the auction could be tailored to raise only as much as necessary to address those concerns (as with RGGI funding of public benefits programs) or made more comprehensive to address credit allocation.

In terms of a long-term price signal, the type of auction employed would have some effect. Likewise, it would have an effect on the stability of that price signal.

An auction could provide substantial incentive for new technology if the auction is structured to encourage a long-term and stable price signal and if revenues received are at least partly directed toward research, development, and demonstration programs.

An auction would probably not reduce the uncertainty with costs, because differing assumptions could be made about the actual operation of the auction, its efficiency, and the effectiveness of the recycled revenues. However, an auction would not have any effect on benefits received by the program, unless it was joined with a safety valve or other limit on auction prices.

Safety Valve

The purpose of a safety valve is to limit the costs of any climate change control program (price) at the potential expense of reductions achieved (quantity). Safety valves encompass a variety of carbon tax-tradeable permit hybrid schemes. Perhaps the most publicized version is that recommended by the National Commission on Energy Policy.[40] The Commission scheme would be implemented through a flexible, market-oriented permit trading program. The total number of permits each year would be based on a mandated decline in GHG intensity and projected GDP growth. However, the scheme includes a cost-limiting safety valve that allows covered entities to make a payment to the government in lieu of reducing emissions. The initial price of such payments would be $7 per ton in 2010. Thus, if a covered entity chooses, it may make payments to the government at a specific price rather than make any necessary emissions reductions.

Effectively, a safety valve places a ceiling on compliance costs — in that way, it acts like a carbon tax. To the extent an entity's control costs, or the permit market, remain below the safety valve, the scheme acts like a tradeable permit program. The degree to which a safety valve reduces costs would depend on the extent to which it is used by entities (e.g., who do not have a cost-effective alternative). However, the complex interactions involved in a scheme that includes both price and quantity controls should not be underestimated. As stated by Jacoby and Ellerman:

> The usefulness of the safety valve depends on the conditions under which it might be introduced. For a time, it might tame an overly stringent emissions target. It also can help control the price volatility during the introduction of gradually tightening one, although permit banking can ultimately serve the same function. It is unlikely to serve as a long-term feature of a cap-and-trade system, however, because of the complexity of coordinating price and quantity instruments and because it will interfere with the development of systems of international emissions trade.[41]

In contrast, Morgenstern argues that the complexity is worth it in preventing price spikes, particularly if a substantial reduction in emissions is envisioned: "If only modest reductions are undertaken, a system of banking and offsets is likely to be adequate in preventing price spikes. In order to achieve more ambitious targets, however, the safety valve is clearly preferred."[42]

To address distributional concerns, a safety value could be tailored for specific sectors to address concerns about cost-effective reduction options or competition. In addition, to the extent the safety valve created revenues, some of the funds raised could be recycled to affected parties.

The effect of a safety valve on new technologies reflects the complexity discussion above. If a low safety valve price was chosen (meaning it would keep compliance costs low), it could have a dampening effect on long-term development of new technology. By creating a ceiling on the value of GHG reductions, but providing no floor for those reductions, a weak market signal may be sent. This might be offset to some degree if funds collected by the safety valve were directed toward new technology, but marketing of any resulting technology might still be difficult if the market price is held low.

A safety valve would dampen the possibility of an upward spike in credit prices — indeed, it is a major reason for considering such an option. However, it would not affect any volatility occurring below the safety valve value and have no effect on a collapse in credit prices. By the same token, the safety valve would put an absolute limit on the projected costs of the program at the level of the safety valve. However, it would do this at the expense of certainty in terms of reductions achieved.

ILLUSTRATIVE APPROACHES

The selected options discussed above are summarized in Appendix A. As suggested, the various options identified have different strengths and weaknesses, depending on the facet of costs one wants to address. Fortunately, many of the options are not mutually exclusive, nor do they require complete adoption — parts of individual approaches can be combined with other parts to meet program specifications in terms of firmness of the goal (also called the "hardness" of the emissions cap) and time frame.

To illustrate, a program focused on achieving a specific tonnage reduction with some flexibility in implementation but not in a manner that threatens the integrity of the cap could incorporate several of these options. The most obvious mechanism to include in the quantity-based cap-and-trade would be banking and borrowing options that would increase flexibility of the program across time without any deterioration in the tonnage requirement. Flexibility and protection against price increases could be enhanced by expanding supply options to include all greenhouse gases, sequestration, and international trading. Depending on one's confidence in the individual supply options, use could be restricted to a maximum percentage of reduction achieved through the option (common in many proposals) or to a more flexible percentage restraint based on credit prices (as proposed by RGGI). Proper monitoring and enforcement could minimize any potential effect on the cap.

This illustration would not necessarily provide either the long-term price signal or funding necessary for new technology. One supplemental option that could help mitigate this problem would be credit auctions. Auctions would have no effect on the cap, but would provide the program with a revenue flow that could be at least partly directed toward research and development. The auction could be designed to raise revenues only by auctioning a small percentage of the credits (such as the current acid rain program), or be comprehensive and auction all credits, thus improving overall economics and providing a clear market signal (as is being proposed by New York to meet its RGGI requirements). In the latter case, coordinating the auction with any trigger price mechanism for expanded supply options would ensure harmonious implementation. Depending on the structure of the auction chosen, the comprehensive auction would also provide a clear market price for reductions and, with the addition of forward markets, some indication of the general direction of those prices.

Finally, the auction and its resulting revenues could also be used to address pressing distributional cost issues. While the mixture of options used in this illustration could potentially mitigate several of the cost issues identified here, it would not provide cost certainty. The quantity side of the equation is the controlling factor under this illustration — prices could be tempered by the market flexibility introduced by the options, but actual costs would not capped.

In contrast, a more price-oriented illustration could employ a safety valve to place an absolute limit on credit prices. In such a hybrid system, the focus of the program is the safety valve limit as much as any tonnage cap. The quantity-based limits of the emissions cap determine the probability that the safety valve would be triggered, assuming a well-functioning market. However, in addition to the supply-demand dynamic that the credit market will reflect, any market failure or disruption resulting from external events could trigger the safety valve for participants. Ultimately, quantity is subordinate to price.

One can potentially reduce the probability that the safety valve would be invoked by including several of the other options discussed here. Expanding supply options would enlarge the pool of available reductions and potentially improve the stability of the credit market if properly integrated. Employing a dynamic tonnage target or an economic-based circuit breaker could help address any economic growth spike that might trigger the safety valve. The question of using these options in a safety valve program is whether they would affect the cap more or less than invoking of the safety valve. In contrast, borrowing and banking would help stabilize markets without having any effect on the cap.

Like the illustration above, this approach would not necessarily promote new technology — indeed the safety valve could discourage such development, unless it generated revenue that was directed toward research and development. If revenues were deemed insufficient for new technology (and to address distributional concerns if desired), the safety valve program could be supplemented with an auction. However, in any case, this illustration is driven by price concerns — concerns that make coordinating new technology development and minimizing impacts on the emission cap difficult.

A final illustration could also be the simplest — imposition of a carbon tax. The clear focus of the program would be the level of the tax, the steepness of any future increases in the tax, and who has the pay the tax. As noted earlier, it could be crafted to address all the cost concerns identified in this report; however, it would represent a new direction in U.S. climate change and current international efforts.

ADDRESSING COSTS THROUGH MARKET MECHANISMS: RESOLVING THE PRICE-QUANTITY ISSUE

The fundamental policy assumption that has changed between the U.S. ratification of the 1992 United Nations Framework Convention on Climate Change (UNFCCC) and the current Bush Administration's decision to abandon the Kyoto Protocol process concerns costs.[43] The ratification of the UNFCCC was based at least partially on the premise that significant reductions could be achieved at little or no cost. This assumption helped to reduce concern some had that the treaty could have deleterious effects on U.S. competitiveness — a significant consideration because developing countries are treated differently from developed countries under the UNFCCC. Further ameliorating this concern, compliance with the treaty was voluntary. While the United States could "aim" to reduce its emissions in line with the UNFCCC goal, if the effort indeed involved substantial costs, the United States could fail to reach the goal (as has happened) without incurring any penalty under the treaty. This flexibility would have been eliminated if the United States had ratified the Kyoto Protocol with its mandatory reduction requirements; the Bush Administration cited this lack of flexibility as a major reason for rejecting the Kyoto process.

With the passage of the 2005 Sense of the Senate climate change resolution calling on Congress to enact a mandatory, market-based program to slow, stop, and reverse the growth of greenhouse gases, the need to address the cost issue has arisen again. Indeed, the resolution itself states that the program should be enacted at a rate and in a manner that "will not significantly harm the United States economy" and "will encourage comparable action" by other nations.[44] Facets of the cost issue that have raised concern include absolute costs to the economy, distribution of costs across industries, competitive impact domestically and internationally, incentives for new technology, and uncertainty about possible costs.

Market-based mechanisms attempt to address the cost issue by introducing flexibility into the implementation process. The cornerstone of that flexibility is permitting sources to decide their appropriate implementation strategy within the parameters of market signals and other incentives. That signal can be as simple as a carbon tax or comprehensive credit auction that tells the emitter the value of any reduction in greenhouse gases, to a credit marketplace that is constrained by a ceiling price (safety valve) and includes incentives for new technology. As illustrated here, the combinations of market mechanisms are numerous, allowing decision makers to tailor the program to address specific concerns.

In a sense, the options discussed here represent a continuum between alternatives focused on the price side of the equation (e.g., carbon taxes) through hybrid schemes (e.g., safety valves) to alternatives focused on the quantity side (e.g., banking and borrowing). They are tools to assist in the assessment of potential greenhouse gas reduction approaches, leaving any policy decision on balancing the price-quantity issue to the ultimate decision makers.

This balance will not be easy to achieve. There is little doubt that market-based mechanisms are available to assist in implementing a GHG reduction program at less cost than more traditional command-and-control methods.[45] Market-based mechanisms offer substantial flexibility to program designers and participants. However, the complexity of market mechanisms (particularly trading programs) increases substantially with the scope of emitting sources included (particularly if international trading is envisioned) and the specificity of any allocation scheme.

Thus, perhaps the most difficult issue to be addressed in designing a market-based implementation strategy for reducing GHG emissions is determining who is included, and who is exempted.

Appendix A: Summary of Selected Options to Address Cost Uncertainty of Greenhouse Gas Reduction Programs

Option	Absolute Costs	Distributional Costs	Long-Term Costs	Price Stability	Cost Uncertainty	Effect on Benefits
Carbon Tax	Allows economics to determine ultimate emission reductions. Costs limited to tax levy. Actual costs would depend on the level of the tax, availability of reduction below that level, and the distribution of the revenues	Distributional concerns about costs can be addressed by either partly or completely exempting specific sectors or targeting sectors with funding from the received tax revenues	Long-term development of new technology would be stimulated by creating a long-term price floor on carbon and strengthened further by targeting R&D with funding from the received tax revenues	Would provide a stable consistent price signal	Would provide an upper limit on potential cost estimates. The lower limit would still be subject to uncertainty	Would make reductions dependent on the level of the tax. The quantity reduction becomes the variable while the price is fixed
Dynamic Tonnage Target	Depending on specifics, would probably offer some cost protection against unforeseen spikes upward in economic growth	Distributional concerns about costs could be addressed by variety of regional or sector-specific, metrics.	Incentive for new technology would depend on the slope of reductions mandated by the program	Would not necessarily avoid short-term fluctuations in market price. Different metrics for different sectors	Would only have modest effect on reducing uncertainty in cost estimates	Depending on the specifics of the target, benefits could be at least slightly dependent on economic conditions
Expanded Supply Options	Can substantially reduce costs, depending on the additional options included.	Can help sectors that do not have cost-effective means of reducing emissions on their own	Depends on how well the additional sources are integrated into the overall market — a stratified market can	Depends on how well the additional sources are integrated into the overall market — a stratified market	Can increase uncertainty by adding new variables to the estimates, including availability,	Should have no effect on reductions achieved, assuming proper safeguards are taken, but new risks are introduced

			muddle the long-term price signal	may result in independent pricing trends	penetration and costs of the additional options	with some options (like international trading)
Economic-Based Circuit Breaker	Reduces costs by temporarily extending compliance deadlines and/or slowing emission reduction targets. The degree of cost savings depends on the specifics of the program	Could address distributional concerns by tailoring its tonnage and timetables triggers to specific sectors	Its short-term focus could muddle the long-term price signal important for developing new technology	Depending on the responsiveness of the tonnage and timetable triggers, it would help mitigate short-term price volatility	Scheme introduces many new uncertainties while reducing others in estimating costs	Increases uncertainty of benefits by making the quantity of reductions achieved a variable
Technology-Based Timetable	Potentially reduces costs by delaying compliance and encouraging more cost-effective approaches in the long-term	Would depend on opportunities for new technologies in given sectors	Arguably, the primary focus of this scheme is to encourage new technology deployment	May have little effect on price stability; indeed, it could increase short-term volatility and uncertainty by removing demand and volume from the market	Scheme introduces new uncertainty to cost estimates	Effect would depend on how widespread the assumed participation rate is.
Banking and Borrowing	Reduces costs by making the emissions cap more flexible over time	Little effect	Can help support a long-term price signal for new technology by supporting prices when costs are lower than expected	The added flexibility can help damp short- term volatility, but not eliminate it	No significant effect	No significant effect over the long-term

Appendix A. (Continued)

Option	Absolute Costs	Distributional Costs	Long-Term Costs	Price Stability	Cost Uncertainty	Effect on Benefits
Auctioning Permits	Allows the marketplace to allocate permits. Actual costs would depend on percentage of permits auctioned and distribution of the revenues	Can be used to address concerns by tailoring auctions for specific sector and/or directing revenues toward affected sectors	Some revenues could be targeted for new technology. Also, auctions would help determine market price of reductions	Depending on the volume of the auction, could have some effect on short- term volatility, but not eliminate it	Scheme introduces new uncertainties to cost estimates	No significant effect
Safety Valve	Effect on cost depends on level that the safety valve is set	Safety valve levels could be tailored for specific sectors	By setting a ceiling, but not a floor on prices, could have a damping effect on new technology depending on the level imposed	Would place an upper limit on price volatility	Would place an upper limit on cost estimates	Would make reductions a function of the safety valve level

ENDNOTES

[1] For an analysis of federal policy and congressional debate since ratification of UNFCCC, see CRS Report RL30024, *Global Climate Change Policy: Cost, Competitiveness and Comprehensiveness*, by Larry Parker and John E. Blodgett.

[2] President George W. Bush, *President Bush's Speech on Global Climate Change* (June 11, 2001).

[3] The six gases recognized under the Kyoto Protocol are carbon dioxide (CO_2), methane (CH_4), nitrous oixde (N_20), sulfur hexafluoride (SF_6), hydrofluorocarbons (HFC), and perfluorocarbons (PFC)

[4] S.Amdt. 866, passed by voice vote after a motion to table, failed 43-54, June 22, 2005.

[5] Finland, the Netherlands, Sweden, Denmark, and Norway.

[6] For CFC-11 and 12, the current (2006) tax is $10.30 per pound. The floor stocks tax is $0.45 per pound (2006). For more specifics on the current tax level, see IRS Form 6627, *Environmental Taxes.*

[7] For example, see Interlaboratory Working Group, *Scenarios for a Clean Energy Future*, ORNL/CON-476 (November 2000).

[8] For example, see *CERA Advisory Service, Design Issues for Market-based Greenhouse Gas Reduction Strategies; Special Report* (February 2006), p. 59; Congressional Budget Office, *Evaluating the Role of Prices and R&D in Reducing Carbon Dioxide Emissions* (September 2006).

[9] Richard D. Morgenstern, *Climate Policy Instruments: The Case for the Safety Valve* (Council on Foreign Relations, Sept. 20-21, 2004), p. 9

[10] For example, see Dallas Burtraw, David A. Evans, Alan Krupnick, Karen Palmer, and Russell Toth, *Economics of Pollution Trading for SO_2 and NOx* (Resources for the Future, March 2005); David Harrison, Jr., *Ex Post Evaluation of the RECLAIM Emissions Trading Program for the Los Angeles Air Basin* (Organization for Economic Co-operation and Development, Jan. 21-22, 2003); and Andrew Aulisi, Alexander E. Farrell, Jonathan Pershing, and Stacy VanDeveer, *Greenhouse Gas Emissions Trading in U.S. States: Observations and Lessons from the OTC NOx Budget Program* (World Resources Institute, 2005).

[11] John P. Weyant, *An Introduction to the Economics of Climate Change Policy* (Pew Center on Global Climate Change, July 2000).

[12] See CRS Report 98-738, *Global Climate Change: Three Policy Perspectives*, by Larry Parker and John Blodgett.

[13] The National Commission on Energy Policy, *Ending the Energy Stalemate: A Bipartisan Strategy to Meet America's Energy Challenges* (December 2005), p. 21.

[14] Vicki Arroyo and Neil Strachan, *Addressing the Costs of Climate Change Mitigation*, presented at the Aspen Workshop: A Climate Policy Framework: Balancing Policy and Politics.

[15] Carbon dioxide (CO_2), methane (CH_4), nitrous oixde (N_20), sulfur hexafluoride (SF6), hydrofluorocarbons (HFC), and perfluorocarbons (PFC).

[16] For more on sequestration approaches, see CRS Report RL33801, *Direct Carbon Sequestering: Capturing and Storing* CO_2, by Peter Folger.

[17] The Regional Greenhouse Gas Initiative (RGGI) is an initiative of currently seven northeastern states to reduce GHG emissions. A signed 2005 memorandum of agreement (MOU) requires the parties to stabilize and then reduce CO_2 emissions from powerplants, implemented through an allowance-based cap-and-trade program. If the allowance price rises above $7, offsets from outside the region may be used for compliance purposes at a 1:1 ratio, with the generator able to cover up to 5% of its emissions. (If below $7, such offsets are discounted 50% and the compliance limit is 3.3% of a generator's emissions. If the allowance price exceeds $10, offsets from international projects could be used to cover up to 20% of a generator's emissions.) For more information, see [http://www.rggi.org].

[18] For a review of these estimates, see CRS Report RL30285, *Global Climate Change: Lowering Cost Estimates through Emissions Trading — Some Dynamics and Pitfalls*, by Larry Parker (available from the author).

[19] For example, see John Reilly, Marcus Sarofim, Sergey Paltsey, and Ronald G. Prinn, *The Role of Non-CO2 Greenhouse Gases in Climate Policy: Analysis Using the MIT IGSM*, MIT Joint Program on the Science and Policy of Global Change, Report No. 114 (August 2004); MIT Joint Program on the Science and Policy of Global Change, *Multi-gas Strategies and the Cost of Kyoto*, Climate Policy Note 3 (April 2000); Vincent Gitz, Jean-Charles Hourcade, and Philippe Ciais, "The Timing of Biological Carbon Sequestration and Carbon Abatement in the Energy Sector Under Optimal Strategies Against Climate Risks," 27 *The Energy Journal* 3 (2006), pp. 113-133.

[20] For more on the ETS, see CRS Report RL33581, *Climate Change: The European Union's Emissions Trading System (EU-ETS)*, by Larry Parker.

[21] Robert Repetto and Duncan Austin, *The Costs of Climate Protection: A Guide for the Perplexed* (World Resources Institute, 1997), p. 21.

[22] CERA Advisory Service, *Design Issues for Market-based Greenhouse Gas Reduction Strategies: Special Report* (February 2006), pp. 54-55.

[23] CERA Advisory Service, *Design Issues for Market-based Greenhouse Gas Reduction Strategies: Special Report* (February 2006), pp. 54-55; Robert Repetto and Duncan Austin, *The Costs of Climate Protection: A Guide for the Perplexed* (World Resources Institute, 1997) p. 22.

[24] M. Jaccard and W.D. Montgomery, "Costs of Reducing Greenhouse Gas Emissions in the USA and Canada," 24 *Energy Policy* 10/11 (1996), pp. 889-898.

[25] The Clean Air Act (42 U.S.C. 7401-7626), Section 110(c)(5)(C).

[26] U.S. Environmental Protection Agency, Letter to California Independent System Operator Corporation (August 12, 2000).

[27] "RECLAIM Poised for Major Changes," *Executive Brief* (New York: Evolution Markets), pp. 1-2.

[28] South Coast Air Quality Management District, Governing Board Meeting (Jan. 7, 2005).

[29] See Clean Power Group website, [http://www.eea-inc.com/cleanpower/index.html].

[30] P.L. 101-559, Title IV, Section 409 (1990).

[31] Dallas Burtraw, David A. Evans, Alan Krupnick, Karen Palmer, and Russelll Toth, *Economics Pollution Trading for SO2 and NOx*, RFF Discussion Paper 05-05 (March 2005), p. 45.

[32] Henry D. Jacoby and A. Denny Ellerman, *The Safety Valve and Climate Policy*, MIT Joint Program on the Science and Policy of Global Change, Report No. 83 (July 2002), p. 9.

[33] Henry D. Jacoby and A Denny Ellerman, *The Safety Valve and Climate Policy*, MIT Joint Program on the Science and Policy of Global Change, Report No. 83 (July 2002), p. 1.

[34] Richard D. Morgenstern, *Climate Policy Instruments: The Case for the Safety Valve*, Council on Foreign Relations (Sept. 2004), p. 10. This option is discussed further below.

[35] See Dallas Burtraw and Karen Palmer, *Summary of the Workshop to Support Implementing the Minimum 25 Percent Public Benefit Allocation in the Regional Greenhouse Gas Initiative*, RFF Discussion Paper DP 06-45 (Sept. 2006).

[36] Used in the SO_2 trading program, credits are allocated gratis to entities in rough proportion to their historic emissions.

[37] Also called an output-based allocation, credits are allocated gratis to entities in proportion to their relative share of total electricity generation in a recent year.

[38] Dallas Burtraw, Karen Palmer, Ranjit Bharvirkar, and Anthony Paul, *The Effect of Allowance Allocation on the Cost of Carbon Emission Trading*, RFF Discussion Paper 01-30 (August 2001).

[39] Burtraw et al., *Allowance Allocation*, p. 30.

[40] The National Commission on Energy Policy, *Ending the Energy Stalemate: A Bipartisan Strategy to Meet America's Energy Challenges* (December 2005).

[41] Jacoby and Ellerman, *The Safety Valve and Climate Policy*, p. 1.

[42] Richard D. Morgenstern, *The Case for the Safety Valve*, p. 10.

[43] For a review of U.S. climate change policy, see CRS Report RL30024, *Global Climate Change Policy: Cost Competitiveness, and Comprehensiveness*, by Larry Parker.

[44] S. Amdt.866, passed by voice vote after a motion to table, failed 43-54, June 22, 2005.

[45] For background on this point with respect to climate change, see CRS Report RL30285, *Global Climate Change: Lowering Cost Estimates through Emissions Trading — Some Dynamics and Pitfalls*, by Larry Parker.

In: Global Climate Change Revisited
Editor: Horace M. Karling, pp. 27-46
ISBN 978-1-59454-039-4

Chapter 2

CLIMATE CHANGE: ACTION BY STATES TO ADDRESS GREENHOUSE GAS EMISSIONS*

Jonathan L. Ramseur

SUMMARY

In the absence of a federal climate change program, a number of states have taken actions that directly address greenhouse gases. States' efforts cover a wide range of policies, from outlining possible strategies to setting mandatory greenhouse gas emission standards. Although much of the early activity was largely symbolic, the more recent state actions have been more pragmatic.

The states' motivations may be as diverse as the actions themselves. Some states are motivated by projections of climatic changes, while others view their policies as economic opportunities. States also point to the potential co-benefits of reducing greenhouse gases: improvements in air quality, traffic congestion, and energy security. Another driver behind state action is the possibility of catalyzing federal legislation.

Most of the states have shown at least a basic interest in climate change issues. Forty-two states have conducted greenhouse gas inventories; 30 states have either completed or are in the process of preparing climate change action plans; 12 states have set statewide greenhouse gas targets. However, only a small number of states have implemented or are creating mandatory emission reduction programs.

The most significant developments have come from California and from a group of states in the Northeast. The Regional Greenhouse Gas Initiative (RGGI), a partnership of seven northeastern states, sets up a cap-and-trade system aimed at limiting carbon dioxide emissions from power plants. This is scheduled to take effect in 2009. California has made several notable steps. In 2004, the state issued regulations to reduce greenhouse gases from motor vehicles. Ten other states have formally adopted California's new vehicle requirements. In 2006, California passed two milestone climate change statutes. The first would establish a

* Excerpted from CRS Report RL33812 dated January 18, 2007.

state-wide cap on greenhouse gases. The second would effectively limit the use of coal-generated electricity in California.

Predicting the precise consequences of the state-led climate change actions is difficult. Some actions, particularly the recent California legislation, may impact energy markets to some degree. Many observers suggest that the quantity and range of state actions will catalyze federal activity. Industry stakeholders are especially concerned that the states will create a patchwork of climate change regulations across the nation. This prospect is causing some industry leaders to call for a federal climate change program. If Congress seeks to establish a federal program, the experiences and lessons learned in the states may be instructive.

Although some states are taking aggressive action, their possible emission reductions may be offset by increased emissions in states without mandatory reduction requirements. This is perhaps the central limitation of state climate change programs in actually affecting total greenhouse gas emissions. Legal challenges represent another obstacle for state programs, particularly for the more aggressive, mandatory programs.

INTRODUCTION

Over the past century, particularly in recent decades, scientists have documented increases in global temperature and sea levels, decreases of sea ice in the Arctic, and melting of continental ice sheets and mountain glaciers. There is increasing evidence that human activities are at least partially responsible for some of these effects.[1] This is based upon the combination of two conclusions. First, global temperature increases are linked in some manner to the measurable increases of greenhouse gas concentrations in the atmosphere.[2] Second, human activities (e.g., fossil fuel combustion, industrial processes, and deforestation) have contributed to the increased concentration of greenhouse gases in the earth's atmosphere.

The link between greenhouse gas emissions and climate change has motivated efforts to achieve reductions of emissions. In 1992, the United States ratified the United Nations' Framework Convention on Climate Change (UNFCCC), which called on industrialized countries to initiate greenhouse gas reduction.[3] However, in early 2001, President George W. Bush rejected the UNFCCC 1997 Kyoto Protocol, which called for legally binding commitments by developed countries to reduce their greenhouse gas emissions.

Over the past decade, the federal government has promulgated or proposed a variety of voluntary and regulatory actions that, while not specifically seeking to reduce greenhouse gases, may have yielded emission reductions as a byproduct.[4] In recent years, there has been some congressional support for a mandatory reduction program. For example, the Senate version of the Energy Policy Act of 2005 included a "sense of the Senate" Resolution[5] stating:

> It is the sense of the Senate that Congress should enact a comprehensive and effective national program of mandatory, market-based limits and incentives on emissions of greenhouse gases that slow, stop, and reverse the growth of such emissions at a rate and in a manner that, No. 1, will not significantly harm the U.S. economy and, No. 2, will encourage other action and key contributors to global emissions.

In the 109th Congress, members introduced several bills that would have established a mandatory greenhouse gas reduction program, but none of the proposals were reported out of committee.[6]

In the absence of action by the federal government to establish a national program that directly addresses greenhouse gas emissions, a number of states (and local governments, whose actions are not covered in this report[7]) have taken action in this arena. States' efforts cover a wide spectrum, from developing climate action plans to setting mandatory greenhouse gas emission standards. While state action is not a new development — some states set greenhouse gas reduction goals as early as 1989, and many states completed action plans in the 1990s — much of the early activity was focused mostly on rhetoric outlining preferable actions rather than on regulatory requirements. However, recent state action has been more significant. Several states now have regulatory programs that limit greenhouse gases from particular sources. A partnership of northeastern states is in the process of setting up a cap-and-trade system aimed at limiting carbon dioxide emissions from power plants.[8] California has recently enacted legislation and regulations limiting greenhouse gas emissions. In 2004, the state issued requirements to reduce greenhouse gases from motor vehicles, a large source of carbon dioxide emissions in the state. In 2006, California passed two significant laws. The first establishes a mandatory state-wide cap on greenhouse gases; the second sets a greenhouse gas performance standard for power plants that applies to both in-state facilities and facilities exporting electricity to California.

The motivating factors for the various states' actions may be as diverse as the actions themselves. Some actions are motivated by projections of climatic changes, such as sea level rise or agricultural impacts. Some states view their greenhouse gas policies as economic opportunities. States want to position themselves for a "less-carbonized" future,[9] by promoting, for example, alternative energy supplies, particularly sources available in-state. Other states champion greenhouse gas reduction policies because of the possible co-benefits: improved air quality, reduced traffic congestion, and less reliance on foreign energy supplies. Another motivating factor for state action is the possibility of catalyzing federal legislation.

This report covers state actions that directly and explicitly address greenhouse gas emissions. First, the report describes the different types of state actions, both individual and cooperative efforts, that are either proposed or underway, and highlights several of the more significant developments. Second, the report examines state actions from a federal policymaking perspective, including both the potential effects of state-led actions and their limitations.

Direct Action versus Indirect Action

Direct state actions that address greenhouse gas emissions include laws, regulations, or policies that are established explicitly to reduce greenhouse gas emissions. In some cases, it is difficult to draw a line between direct and indirect actions, because a specific policy may be undertaken for multiple purposes, including greenhouse gas reduction. One of the best examples of this ambiguity is a Renewable Portfolio Standard (RPS). An RPS requires that a certain amount or percentage of electricity is generated from renewable energy resources (e.g., solar, biomass). Twenty-one states have implemented an RPS.[10] Although greenhouse gas reduction is not the primary driver for an RPS in most states, some states list their RPS as part of a comprehensive strategy to reduce greenhouse gases.

Indirect actions are often characterized as "no regrets" approaches, providing net benefits regardless of the magnitude of their impacts on climate change. For the purposes of this report, indirect actions are those developed primarily to address other concerns, such as improvements in energy efficiency, energy security, or air quality. Examples of indirect actions include:

- Building codes: A majority of states have building codes that promote energy efficiency in commercial and residential structures.[11]
- Appliance Standards: Twelve states have set energy efficiency standards for appliances that are not covered under the federal program.[12]
- Agricultural policies: Several states promote agricultural practices that may indirectly reduce greenhouse gas emissions. For example, a "no-till" farming technique saves fuel and man hours, while keeping carbon stored in the soil.[13]

This report, however, does not attempt to discuss the extremely wide variety of such indirect actions.

Direct Actions by States

States are implementing a range of direct actions to address greenhouse gas emissions. States' efforts have progressed recently in both quantity and substance. Arguably, early state actions were largely symbolic. In the late 1980s, Vermont[14] and Oregon[15] were the first states to set greenhouse gas reductions goals, but during the subsequent decade (1990-2001), both states increased their greenhouse gas emissions: Vermont by 18% and Oregon by 30%.[16] However, a majority of states have more recently begun to develop their own climate change strategies or policies, with a small but increasing number of states adopting or proposing more significant provisions, including mandatory greenhouse gas reductions.

States have developed and are crafting climate change policies both individually and in cooperation with other states. This section describes the spectrum of direct state actions, identifies the level of participation in various activities, and highlights individual and cooperative state programs when appropriate.

State Action Plans

Thirty states have either completed or are in the process of preparing climate change action plans (see Figure 1). Typically, state action plans are drafted by a climate change task force, composed of members with diverse backgrounds and expertise. In general, task force members examine their state's sources of greenhouse gases, and identify and rank the policy options that are most appropriate (i.e., cost-effective, politically feasible, etc.) for controlling emissions in their state. Often the state action plan is made available for public comment, revised if necessary, and then submitted for approval to state officials.

Reflecting the fact that states have different economic sectors, natural resources, and political structures, state climate change action plans can vary substantially. Some state action plans focus more on indirect, "no regrets" strategies, such as improved energy efficiency, which will likely yield benefits irrespective of climate change effects. Other state action plans are more comprehensive and recommend a portfolio of direct efforts that address greenhouse gases. Although the state climate change action plans may *recommend* an array of policy options, the plans do not necessarily result in direct actions to reduce greenhouse gases. However, the number of completed state plans indicates the interest that a majority of states have in addressing climate change mitigation on some level.

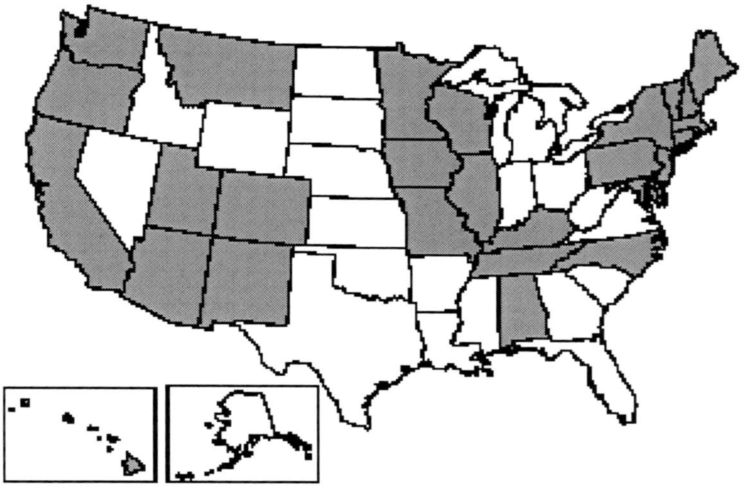

Figure 1. States with Climate Change Action Plans.

Greenhouse Gas Emissions Targets

State emissions targets are goals by which a state can measure its progress in achieving greenhouse gas emissions reduction. By themselves, state emissions targets do not directly reduce greenhouse gases, but a target signals that state officials consider climate change an important issue.

Twelve states have established *statewide* targets for greenhouse gas emissions (see Table 1).[17] Other than California's 2020 target (discussed later), the statewide targets do not constitute legally binding requirements. Nearly all of these states are either in the Northeast or on the west coast of the United States. The New England states' targets are identical, because they are part of a cooperative plan developed in 2001.[18] Of the 12 states in Table 1, New Mexico stands out, because it is the first state with substantial coal and petroleum resources to set an overall emissions target.[19]

Considering the greenhouse gas limits and targets set on the international stage in past years, these state targets are relatively modest. Only New Jersey's greenhouse gas target — 3.5% below 1990 levels by 2005 — was comparable to the U.S. target set by the Kyoto Protocol.[20] However, New Jersey did not meet its 2005 target.[21]

Table 1. Statewide Greenhouse Gas Targets Compared with Emissions Data from 1990 and Recent Years

State	Greenhouse Gases Target	Greenhouse Gas Emissions (million metric tons of CO2 equivalent)				Carbon Dioxide Emissions from Fossil Fuel Combustion (millionmetric tons)			
		1990	1999	2000	2001	1990	2001	2002	2003
AZ[22]	2000 levels by2020	70	87	91	94	62	87	87	88
CA[23]	2000 levels by 2010; 1990 levels by 2020	415	417	434	451	359	383	380	384
CT[24]	1990 levels by2010; 10% below 1990 by2020	43	44	45	44	41	41	40	42
ME[25]	1990 levels by 2010; 10%below 1990 by2020	21	21	23	23	19	22	23	23
MA[26]	1990 levels by 2010; 10%below 1990 by2020	89	85	86	87	83	81	82	86
NH[27]	1990 levels by 2010; 10% below 1990 by2020	16	18	18	18	15	17	17	20
NJ[28]	3.5% below1990 by 2005	123	132	133	132	114	123	123	124
NM[29]	2000 levels by 2012; 10% below 2000 by2020	59	64	66	67	52	57	55	57
NY[30]	5% below 1990by 2010; 10%below 1990 by2020	234	236	239	239	208	207	200	214
OR[31]	Stabilize by 2010; 10%below 1990 by2020	39	52	50	51	31	42	40	40
RI[32]	1990 levels by 2010; 10%below 1990 by2020	10	14	12	13	9	12	12	11
VT[33]	1990 levels by 2010; 10% below 1990 by2020	7	8	8	8	5	7	6	6

Source: Prepared by the CRS with data from the following: state targets compiled by Pew Center on Global Climate Change, at [http://www.pewclimate.org]; greenhouse gas emissions data from World Resources Institute, Climate Analysis Indicators Tool, at [http://cait.wri.org/] (this website notes that the greenhouse gas data excludes land use changes); carbon dioxide data from EPA's Climate Change, State Emissions website, at [http://epa.gov/climatechange/emissions/state.html].

Note: The states' targets are for all greenhouse gases. Carbon dioxide emissions from fossil fuel combustion are also included, because (1) this is the primary source of greenhouse gases in most states and (2) the carbon dioxide data provide two additional years of information.

Table 1 compares the states' emissions in 1990 with emissions from the most recent three years of available data (overall greenhouse gases and carbon dioxide). The emissions data show the reductions states would need to make to meet their established targets. Although some of the 12 states appear within reach of their 2010 targets, the most recent data from many of these states suggest that emissions are not decreasing, but at best are leveling off. More years of data are needed to evaluate progress, primarily because many of the states issued their greenhouse gas targets after 2003, and state-level data after 2003 are not yet available. Moreover, the emissions targets were typically created in conjunction with greenhouse gas reduction policies — some of them mandatory limits on specific industries or segments of state activities — whose implementation is not reflected in the available emissions data.

Greenhouse Gas Emissions Tracking

Reliable greenhouse gas emissions data are a keystone component of any climate change program. To implement effective solutions to climate change, policymakers need up-to-date and accurate information detailing the volume and sources of greenhouse gases in their states. Precise monitoring is particularly vital for market-oriented approaches to greenhouse gas control. Whether a market-oriented program is based on tradeable emissions credits or a carbon tax, reliable and transparent emissions data would be the foundation for developing the allocation systems, reduction targets, and enforcement provisions.

The federal government has several programs in place that either track or estimate greenhouse gas emissions:

- Power plants subject to the 1990 Clean Air Act acid rain program must monitor and report to EPA various air pollutants, including carbon dioxide.[34]
- The Department of Energy administers a voluntary greenhouse gas reduction registry. This program started in 1994, pursuant to Section 1605(b) of the Energy Policy Act of 1992 (P.L. 102-486).[35]
- The EPA prepares an annual inventory of the nation's greenhouse gas emissions and sinks, which is submitted to the United Nations in accordance with the Framework Convention on Climate Change.

Many states have developed, or begun to develop, their own greenhouse gas tracking programs. Although tracking programs may overlap in purpose and terminology, for this report, tracking programs are divided into three categories: inventories, registries, and mandatory reporting.

Greenhouse Gas Inventories

Forty-two states have developed greenhouse gas inventories. Inventories typically provide estimates of emissions for various categories: economic sector (e.g., energy, agriculture), emissions source (e.g., automobiles, power plants), greenhouse gases (e.g., carbon dioxide, methane). In general, states create their inventories by following guidelines developed by the Environmental Protection Agency (EPA) that are based on internationally recognized standards. Inventories are often used to obtain an overall assessment of a state's emissions levels and sources, and are perhaps best suited for monitoring trends and/or developing comprehensive strategies. Although some states have performed inventory updates, most of the states' inventories only cover 1990 emission levels.

Greenhouse Gas Registries

A state greenhouse gas registry is a further step in greenhouse gas tracking. In general, state greenhouse gas registries are voluntary programs that allow facilities to submit and officially record emissions data. Five states have passed legislation to establish greenhouse gas registries, of which three are now underway:[36]

- New Hampshire: The New Hampshire Greenhouse Gas Registry went into effect in 2001. The registry is intended to record emissions reductions in a state database that can be used in addressing possible future requirements.
- California: The California Climate Action Registry began operations in 2002. This state registry is arguably the most comprehensive, as participants register all of their GHG emissions for operations in California; other state (and federal) registries cover only emission reductions. The registry has over 100 participants.
- Wisconsin: The Wisconsin Voluntary Emission Reduction Registry, a registry of voluntary reductions of greenhouse gases, went online in 2003.

Other states are joining forces to establish regional registries. Ten New England and Mid-Atlantic states are developing the Eastern Climate Registry.[37] In addition, the Lake Michigan Air Directors Consortium (LADCO) is working on a registry for several states in the Midwest.[38]

The states' voluntary registry programs encourage participation through incentives. Perhaps the primary incentive is the opportunity for participants to create an official record of emissions reductions, which the parties hope will count as emissions credits in future mandatory reduction programs. At a minimum, participants typically receive some public recognition for their efforts, which may help promote a company's environmental stewardship profile.

Mandatory Greenhouse Gas Reporting

Mandatory reporting programs allow states to monitor greenhouse gas emissions from precise sources. Although the primary purpose of mandatory reporting is typically to support an emission reduction program, a reporting program can potentially provide benefits without an accompanying reduction requirement. For example, if companies' greenhouse gas emissions were made publicly-available and thus comparable, the companies might have an incentive to reduce emissions voluntarily.[39] However, there is some concern that emissions may increase under a mandatory reporting program, especially if companies suspect that the state will establish a mandatory reduction regime in later years. For instance, facilities may attempt to "game" the system by deliberately increasing emissions (or over-reporting them) in order to gain additional allowances once a reduction program is established.

A few states already require, and others are in the process of developing, greenhouse gas emissions reporting as part of an emissions reduction program (discussed in the next section).

Four states currently have a mandatory reporting program that is not linked with an emissions reduction requirement:

- Wisconsin: In 1993, the state established a mandatory reporting program that includes carbon dioxide reporting for facilities generating over 100,000 tons annually.[40]
- New Jersey: Certain facilities in New Jersey that report air pollutant emissions must also submit emission data for carbon dioxide and methane. This requirement went into effect in 2003.[41]
- Maine: Facilities in Maine that emit any criteria pollutant over a specific reporting threshold must also report greenhouse gas emissions. This provision went into effect July 2004.[42]
- Connecticut: Starting in 2006, facilities subject to federal reporting under Title V of the Clean Air Act must submit greenhouse gas emissions data on an annual basis.[43]

Mandatory Programs to Reduce Greenhouse Gases

Mandatory programs to require greenhouse gas reductions represent the most aggressive end of the state action spectrum.[44] As with state actions overall, these programs can vary significantly in scope, stringency, and design. Mandatory programs are generating considerable interest and some controversy. This section discusses the different types of mandatory programs and highlights particular state actions that are currently in effect or under development.

Emission Reduction from Power Plants

A sector-specific approach that focuses on carbon dioxide is relatively easier to implement than an economy-wide program that includes multiple greenhouse gases. The electricity-generating sector is often considered a primary candidate for emission reduction, because in most states electric power plants account for the highest percentage of carbon dioxide emissions.[45] Many of these facilities are already tracking their carbon dioxide emissions as required by the 1990 Clean Air Act.

Regional Greenhouse Gas Initiative

One of the more significant climate change developments at the state level is the Regional Greenhouse Gas Initiative (RGGI). RGGI is a market-oriented effort of seven states — Connecticut, Delaware, Maine, New Hampshire, New Jersey, New York, and Vermont — to reduce carbon dioxide emissions from power plants. RGGI would set up the nation's first mandatory cap-and-trade program for carbon dioxide. The initial objective of RGGI is to stabilize current carbon dioxide emissions from power plants in RGGI states, starting in January 2009, followed by a 10% reduction by 2019. A primary strategy of RGGI is to create a program with flexibility, so that in the future other emission sources/sectors, greenhouse gases, or states could be included. Maryland and Massachusetts are expected to join RGGI in 2007.

Some observers consider RGGI to be a possible test-case for a federal cap-and-trade program, and thus several of RGGI's design elements are generating interest and debate. For example, one specific feature — the emission allocation scheme —is drawing both praise and criticism. In both RGGI's Memorandum of Understanding and its Model Rule, states agreed that at least 25% of emission allowances will be allocated for a "consumer benefit or strategic energy purpose."[46] Some states (e.g., Vermont, New York) have indicated that they intend to allocate 100% of their states' allowances for that purpose. This action would require power plants to purchase the set-aside allowances, most likely through an auction, instead of receiving them at no charge.[47]

Although RGGI is one of the more aggressive state programs addressing climate change, the program will likely face several obstacles. For example, RGGI proponents expect the program to face legal challenges, which could delay program initiation. In addition, a critical design detail — electricity imports from non-RGGI states — is unresolved. This is often described as the "leakage" problem. Leakage can occur when an emissions reduction program does not include all sources contributing to the environmental problem. For example, if a RGGI state lowers its emissions by importing more power from a non-RGGI state, the emissions reductions in the RGGI state may be offset by an emission increase in the exporting state.

Individual State Efforts

Two states have already established emission reduction requirements at *existing* power plants:

- Massachusetts: In 2001, Massachusetts became the first state to take formal action on carbon dioxide emissions at operational power plants. As part of a multi-pollutant strategy, which went into effect in 2006, the state's six largest power plants must reduce carbon dioxide to levels consistent with those produced in the late 1990s. In 2008, this cap is lowered further.[48] The program allows the plants to either make the reductions, demonstrate offsite reductions, or purchase emissions credits from other verifiable sources.
- New Hampshire: In 2002, the state enacted multi-pollutant legislation[49] requiring its three fossil fuel power plants to reduce carbon dioxide to 1990 levels by the end of 2006. In order the meet the cap, the law allows sources to bank early reductions or buy credits through other programs deemed acceptable by state officials.

Both Oregon and Washington have programs that require *new* power plants to reduce carbon dioxide emissions or purchase offsets. In 1997, Oregon became the first state to regulate carbon dioxide emissions by passing legislation[50] requiring new power plants to equal or exceed carbon dioxide levels that are 17% below the best natural gas-fired plant in the nation. Plants can either reduce emissions directly or purchase offsets from a nonprofit organization (the Oregon Climate Trust) that was established with the 1997 law. This organization helps develop various projects that will reduce or sequester greenhouse gases. These projects generate the pool of offsets available (by purchase) to the power plants. So far, all of the new facilities have chosen to purchase offsets instead of reducing onsite emissions.[51] Washington passed similar legislation in 2004, requiring new power plants to offset 20% of their carbon dioxide emissions.[52]

Emission Reductions from Motor Vehicles

The U.S. transportation sector accounts for a substantial percentage — 33% in 2005[53] — of the nation's carbon dioxide emissions. In 14 states, the transportation sector is the single largest source of carbon dioxide emissions.[54] California's transportation sector, in particular, generates almost 60% of the state's carbon dioxide emissions.[55]

California is in a unique position regarding the regulation of air emissions from motor vehicles. It is the only state with conditional authority (i.e., the state needs a waiver from EPA) to develop motor vehicle pollution standards that are more stringent than federal requirements.[56] The law permits other states to choose to follow California's more stringent provisions,[57] and many states have adopted California standards in the past.

In 2002, California enacted the first state law (AB 1493) requiring greenhouse gas limits from motor vehicles.[58] As directed by the statute, the California Air Resources Board (CARB) issued regulations in September 2004, limiting the "fleet average greenhouse gas exhaust mass emission values from passenger cars, light-duty trucks, and medium-duty passenger vehicles."[59] The fleet average caps first apply to model year 2009 vehicles. The caps become more stringent annually, so that by 2016, the fleet average would be 30% below the 2009 level.

At least 10 states (Oregon, Washington, and eight states in the Northeast) have formally adopted the California regulation. However, whether the more stringent standard goes into effect remains uncertain. California requested a waiver (as required by Section 209 of the Clean Air Act) in December 2005, but EPA has yet to respond. Although EPA has approved every California waiver request since 1975, this waiver request may not follow the same pattern. EPA has generally displayed a resistance to use the Clean Air Act to control greenhouse gases.[60] A second hurdle involves several legal challenges from automotive industry groups. Car dealers and trade associations have filed suits in California, Vermont, and Rhode Island, seeking to halt the regulations on various grounds. For example, the plaintiffs contend that California's regulations are preempted by the Energy Policy and Conservation Act (P.L. 94-163), which directs states not to regulate fuel economy standards. For more discussion concerning these legal issues, see CRS Report RL32764, *Global Warming: The Litigation Heats Up*, by Robert Meltz.

California's Statewide Emission Reductions Law

In September 2006, California enacted landmark legislation that would establish a comprehensive greenhouse gas reduction regime. The legislation — AB 32 or the Global Warming Solutions Act[61] — directs the California Air Resources Board (CARB) to develop and implement a statewide program that would reduce the state's greenhouse gas emissions to 1990 levels by 2020.

The statute grants considerable authority to CARB, which is charged with determining critical details concerning the framework and applicability of the program. For example, the law does not specifically require the use of a market-based system, such as a cap-and-trade program, to reduce greenhouse gases. Instead, AB 32 authorizes CARB to develop regulations to "achieve the maximum technologically feasible and cost-effective greenhouse gas emission reductions...." Moreover, the statute does not include a list of regulated emission sources or categories,[62] but instructs CARB to determine which sources are necessary to meet the statewide target.[63]

The law establishes a schedule for various agency deadlines. By June 30, 2007, AB 32 instructs CARB to identify the early reduction options, which can be implemented prior to the mandatory program, and for which a facility will receive emissions credit. The law requires CARB to set up a mandatory reporting scheme by January 1, 2008. Data from the reporting program will be used to establish baselines for emissions sources, which will be subject to emission reductions starting in 2012.

The statute requires the program to account for greenhouse gas emissions from all electricity consumed in California. The agency will need to count emissions connected with electricity that is generated from outside the state. This provision is significant, because it addresses the "leakage" issue. Without this provision, California utilities might have a financial incentive to import more electricity from out-of-state generators, who are not subject to the cap. In such a scenario, California emissions would decrease, but the benefit would be negated by increased emissions in neighboring states.

When developing the emission reduction program in California, AB 32 instructs CARB to consider other greenhouse gas reduction regimes, including RGGI and the European Union's emission trading program. This instruction might open the door for future emissions trading between California and other states.

Other Mandatory Programs

Although they do not require emission reductions or offsets from specific facilities or sources, other mandatory programs may have an impact on greenhouse gas emissions. A few states, California in particular, have recently developed requirements that aim to influence investment in long-term power generation. These state actions may impact greenhouse gas levels by influencing which energy sources — coal, oil, natural gas, etc. — are used to generate electricity for consumers.

California's Greenhouse Gas Emissions Performance Standard

The most significant state action in this regard is California's greenhouse gas performance standard. In September 2006, the state passed legislation that will forbid "load-serving entities"[64] from entering into new "long-term financial commitments"[65] with power plants unless the plant's greenhouse gas emissions are as low or lower than those of a new, combined-cycle natural gas facility.[66] This emissions performance standard will apply to both in-state power plants and out-of-state facilities that seek to export electricity to California. The law directs the California Public Utilities Commission (PUC) to develop the performance standard by February 1, 2007.

Once the new performance standard is applicable (and previous commitments expire), it will effectively prohibit California consumers from using electricity generated by conventional coal-fired power plants. Compared with a combined-cycle natural gas plant, a conventional coal-fired power plant emits more than twice the amount of carbon dioxide. Using current technologies, coal-fired generators would fail to meet the new emissions standard.[67] As the law takes effect, California will likely need to replace its coal-generated electricity with alternative sources of power.

The new emissions standards will impact not only California, but also other states in the West. Although California's electricity imports generally fall between 22% and 32% of the state's total electricity consumption, its imports are responsible for 39% to 57% of the total greenhouse gas emissions linked with electricity.[68] This is due to the fact that most of California's in-state electricity is produced from sources other than coal, while most of the state's imported electricity is generated through coal combustion. Once the standard takes effect, the coal-fired plants in neighboring states, which previously provided electricity to California, will need to look elsewhere for customers. The same goes for coal-fired power plants still in development in western states, which may have been designed, at least in part, to serve California consumers.[69]

Greenhouse Gas "Adders"

Another state action that may affect a state's sources of electricity generation is the adoption of a greenhouse gas (or carbon) adder. In general, adders require utilities to weigh the future costs of greenhouse gas emissions when considering different energy investment options (e.g., fossil fuels, renewable energy supplies). For example, California's Public Utilities Commission requires investor-owned-utilities to include a value of $8/ton of carbon dioxide emissions when conducting long-term planning or procurement activities.[70] The agency stated that this requirement "will serve to internalize the significant and under-recognized cost of [greenhouse gas] emissions, [and] help protect customers from the financial risk of future climate regulation...."[71] Only a few other states[72] require some type of greenhouse gas adder, and California's adder may be rendered less relevant due to its new emission performance standard (discussed above). At this stage, the adders have not been credited with changing any procurement decisions.[73]

Issues for Congress

The climate change activity in the states raises several issues that may be of interest to Congress. This section discusses some of the potential effects of state action in lieu of federal legislation. This section also examines the limitations of state actions, both from a climate change policy perspective and in the context of legal challenges.

Potential Effects of State Actions

Many states generate significant emissions of greenhouse gases. If individual U.S. states were classified as sovereign nations, 21 states would rank in the top 60 for nations that annually emit the most carbon dioxide.[74] Compared with other nations, Texas, the combined RGGI states, and California rank as top carbon dioxide emitters (see Table 2).

Although the states developing mandatory reduction programs — California and the RGGI participants — account for an appreciable percentage of U.S. carbon dioxide emissions (almost 20%), most of the states are pursuing considerably less aggressive climate change policies. Unless the more aggressive state actions foster greater participation or catalyze a federal program, the current state actions are unlikely to impact global climate change. With this range of state activity, it is difficult to predict the precise consequences of state-led climate change actions. This section highlights possible effects from state actions.

Table 2. Top Carbon Dioxide Emissions by Nation and U.S. States (2001 data)

Country, State, or Group	Carbon Dioxide Emissions (million metric tons)
United States	5,728
European Union	3,928
China	3,452
Russian Federation	1,544
Japan	1,221
India	1,068
Germany	884
Texas	*678*
RGGI states[75]	*594*
United Kingdom	562
Canada	522
South Korea	473
Italy	448
France	389
Mexico	388
California	*386*

Source: Prepared by CRS with data from World Resources Institute, Climate Analysis Indicators Tool, at [http://cait.wri.org/] Note that the carbon dioxide data excludes land use changes.

States as Policy Laboratories

A central argument in support of state climate change action is that states can serve as laboratories for policymaking. States can test different ideas and policies on a smaller scale, and help determine which climate change solutions are most effective. For example, there has been some debate regarding how a cap-and-trade program might work on a national level. Although the federal acid rain program, which involves sulfur dioxide emissions trading, is generally considered a success, emissions trading programs for other purposes have encountered problems during implementation. State programs offer the opportunity to iron out logistical details that are crucial in a cap-and-trade system: for example, which sources to regulate; how to allocate emissions allowances; how high to set the emissions cap; when to allow offsets instead of actual reductions.

State programs can inform federal policymakers in other ways. The political process by which states create climate change policy can be enlightening and perhaps adaptable on the federal level. For instance, by examining the development and passage of state legislation, federal policymakers may better understand the motivations of different stakeholders and learn how best to frame the issues.

Possible Market Influences

Mandatory emission reduction programs may have some effect on energy markets. For example, California's recently enacted greenhouse gas performance standards, once underway, will effectively bar California consumers from using electricity generated by conventional coal-fired power plants. In general, states with emission reduction requirements might see an increase in the use and support of less carbon-intensive fuels. This increase in carbon-regulated states may shift more carbon-intensive fuel use to states without emissions regulations. The increased supply of more carbon-intensive fuels (primarily coal) could result in lower electricity prices in states without greenhouse gas emissions requirements. Federal policymakers may consider whether these possible outcomes coincide with federal energy priorities.

Patchwork of Regulations

One concern shared by many observers, particularly industry stakeholders, is that state climate change programs (in lieu of a federal program) will create a patchwork of regulations across the nation. A patchwork system of standards may hinder a company's efficiency and possibly create economic burdens for firms that operate in multiple states. The prospect of regulations that vary from state to state is driving some companies to support a federal climate change program with comparable requirements across the entire United States.

LIMITATIONS OF STATE ACTIONS

Climate change has been described as the "ultimate global commons problem."[76] The global warming and climate impacts associated with increased greenhouse gases in the atmosphere cannot be linked with specific emission sources. Unlike localized reductions in other air pollutants (e.g., sulfur dioxide, particulate matter), when an emissions source reduces its carbon dioxide emissions, it does not generate a corresponding local climate change benefit unless there are similar widespread reductions globally or at least in wide areas.

From a practical standpoint, the actions of one or a group of states or nations cannot by themselves reduce the global accumulation of greenhouse gases in the atmosphere. However, as discussed above, actions now underway by many states in the United States may create examples and/or models that will prove instructive in more widespread applications. Moreover, when business and industry have confronted a growing patchwork of state requirements, these sectors have historically begun to favor a national policy — as has begun to happen in the case of state-level actions on climate change. However, the lack of a national program or a truly global approach to greenhouse gas emissions reductions does limit what individual states can accomplish in actually reducing greenhouse gas emissions and accumulations.

Legal challenges may further limit the effectiveness of state action. The possibility of legal challenges creates considerable uncertainty regarding the future of state climate change actions, particularly the more progressive programs. There are already several lawsuits (discussed above) against state actions that seek to regulate greenhouse gas emissions from motor vehicles. Further litigation confronting other types of state action is anticipated. For example, many expect a legal challenge against the RGGI program when the first state's rule is officially issued.[77] There is some question as to whether California's recently enacted greenhouse gas performance standards are constitutional.[78] Arguably, the standards disproportionately impact the neighboring states that have historically exported coal-generated electricity to California consumers. The legal arguments in these cases are beyond the scope of this report, but many observers conclude that it is difficult to predict how the courts will interpret and decide upon these issues.

ENDNOTES

[1] This report does not address the debates associated with the climate change science nor the role of human activity.

[2] For example, carbon dioxide, the primary greenhouse gas, has risen worldwide from 280 parts per million (ppm) to over 380 ppm over the past 150 years.

[3] The United Nations Framework Convention on Climate Change (UNFCCC) defines greenhouse gases to include carbon dioxide, methane, nitrous oxide, hydrofluorocarbons, perfluorocarbons, and sulfur hexafluorane.

[4] For example, federal programs that promote energy efficiency or the use of renewable energy sources have the potential to reduce greenhouse gas emissions.

[5] Senate Amendment No. 866 to H.R. 6, passed by voice vote June 22, 2005. A motion to table the amendment was rejected by a roll call vote (44 - 53).

[6] For a more thorough discussion regarding federal climate change legislation and policy, see CRS Report RL31931, *Climate Change: Federal Laws and Policies Related to Greenhouse Gas Reductions*, by Brent D. Yacobucci and Larry Parker, and CRS Report RL32955, *Climate Change Legislation in the 109th Congress*, by Brent D. Yacobucci.

[7] A number of local governments are pursuing activities that may directly or indirectly reduce greenhouse gas emissions. For example, 171 local governments (cities, counties) in 35 states have joined the Cities for Climate Protection (CCP). Participating entities commit to reduce local emissions that contribute to global warming. For more information on this program, see [http://www.iclei.org/index.php?id=1118].

[8] In a cap-and-trade system, regulators set a cap (or limit) on the overall emissions of a given gas from a specified group of sources, such as power plants. The emissions allowed under the new cap are then allocated in the form of credits (or permits) to individual sources. Sources that emit more than their allowance must buy credits from those who emit less than their allowance, thus creating a financial incentive for sources to reduce their own emissions. For more information on cap-and-trade systems, see EPA's website at [http://www.epa.gov/airmarkets/capandtrade].

[9] See Rabe, Barry, 2006, "Second Generation Climate Policies in American States: Proliferation, Diffusion, and Regionalization," *Issues in Governance Studies*, The Brookings Institution, August 2006.

[10] Two additional states are considering an RPS. See EPA, Summary of State Clean Energy-Environment Policy Data Table (current as of 11/21/2005), at [http://www.epa.gov/cleanenergy/stateandlocal/activities.htm]. See also the Pew Center on Global Climate Change, Map: States with Renewable Portfolio Standards, at [http://www.pewclimate.org]. This map identifies Illinois and the District of Columbia as also having established an RPS.

[11] EPA data indicates 35 states have commercial codes and 31 states have residential codes for energy efficiency. See EPA, Summary of State Clean Energy-Environment Policy Data Table (current as of 11/21/2005).

[12] See EPA, Map: State Energy Efficiency Actions - State Appliance Efficiency Standards (as of 6/23/06), at [http://www.epa.gov/cleanenergy/stateandlocal/activities.htm].

[13] Georgia promotes this technique through its No-Tillage Assistance Program (NTAP), which provides equipment and funding assistance. See Pew Center on Global Climate Change, State and Local Net Greenhouse Gas Emissions Reduction Programs, at [http://www.pewclimate.org].

[14] Vermont Executive Order 79 (October 23, 1989) called for a 15% reduction below 1989 levels by 2000. See U.S. Congress, Office of Technology Assessment, 1991, *Changing by Degrees: Steps to Reduce Greenhouse Gases*, p. 327.

[15] Oregon Senate Bill 576 (1989) set a goal of 20% reduction of 1988 levels by 2005. See U.S. Congress, Office of Technology Assessment, 1991, *Changing by Degrees: Steps to Reduce Greenhouse Gases*, p. 327.

[16] See World Resources Institute, Climate Analysis Indicators Tool, at [http://cait.wri.org/].

[17] Several states have also developed more narrow targets, either for industry or electricity generation or only for carbon dioxide emissions.

[18] New England Governors/Eastern Canadian Premiers, *Climate Change Action Plan 2001*, August 2001, at [http://www.negc.org].

[19] In 2005, New Mexico ranked 11th in coal production and 6th in crude oil production. The state also ranked 3rd in natural gas production, a fuel that releases significantly less greenhouse gas than coal or oil when burned. See U.S. Department of Energy, Energy Information Administration Statistics, at [http://www.eia.doe.gov/].

[20] The U.S. Kyoto target was 7% below 1990 levels, averaged over the commitment period 2008 to 2012. For more on international climate agreements and U.S. involvement, see CRS Report RL33602, *Global Climate Change: Major Scientific and Policy Issues*.

[21] Per telephone discussion with New Jersey state official (January 5, 2007).

[22] Arizona Executive Order 2006-13 (September 7, 2006).

[23] California Executive Order S-3-05 (June 1, 2005) set the 2010 and 2020 targets; AB 32 (discussed below) made the 2020 target mandatory.

[24] Connecticut Public Act No. 04-252 (June 14, 2004).

[25] Maine LD 845 (HP 622) (effective September 13, 2003).

[26] Massachusetts Climate Protection Plan of 2004 (Spring, 2004).

[27] The Climate Change Challenge (December 2001).

[28] New Jersey Administrative Order 1998-09 (March 17, 1998).

[29] New Mexico Executive Order 05-033 (June 9, 2005).

[30] New York State Energy Plan (June 2002).

[31] Oregon Strategy for Greenhouse Gas Reductions (December 2004).

[32] Rhode Island Greenhouse Gas Action Plan (July 2002).

[33] This target is discussed in Vermont's state plan, Fueling Vermont's Future: Vermont Comprehensive Energy Plan and Vermont Greenhouse Gas Action Plan (July 1998).

[34] Section 821, 1990 Clean Air Act Amendments (P.L. 101-549, 42 USC 7651k). For more information regarding federal programs see CRS Report RL31931, *Climate Change: Federal Laws and Policies Related to Greenhouse Gas Reductions*, by Brent D. Yacobucci and Larry Parker.

[35] For more information on this program, see [http://www.eia.doe.gov/oiaf/1605/frntvrgg.html].

[36] The other two states are Maine and Georgia. Maine's registry is not yet operational.

Georgia, instead of tracking greenhouse gas emissions, established a registry for counting the offsetting reductions in greenhouse gases obtained by carbon sequestration. Not counted as one of the five states, New Jersey repealed a previously enacted registry program in 2004.

[37] These states include Connecticut, Delaware, Maine, Massachusetts, New York, New Hampshire, New Jersey, Pennsylvania, Rhode Island, and Vermont. For more information on this partnership, see [http://www.easternclimateregistry.org/].

[38] Illinois, Indiana, Michigan, Ohio, and Wisconsin. For more on this registry, see [http://www.ladco.org/regional_greenhouse.htm].

[39] This notion is analogous to the arguments in support of EPA's Toxic Release Inventory (TRI) Program, which requires facilities to submit annual data concerning their releases of chemicals to the environment. The TRI program is generally considered a success, as releases have decreased since the program's inception. Rabe, Barry, 2002, *Greenhouse & Statehouse: The Evolving State Government Role in Climate Change*, Prepared for the Pew Center on Global Climate Change.

[40] Wisconsin Chapter NR 438.03.

[41] New Jersey Administrative Code 7:27-21.3.

[42] Maine Department of Environmental Protection Rules, Chapter 137 (per 38 MRSA, Section 575).

[43] Connecticut Public Act No. 04-252 (June 14, 2004).

[44] Several states have voluntary reduction programs. Most of these were discussed earlier in the context of state emissions registries.

[45] Based on 2001 data. Energy Information Administration, Emissions of Greenhouse Gases in the United States 2004 (Table C2).

[46] See RGGI Model Rule, issued August 15, 2006, p. 42; and RGGI Memorandum of Understanding, Section G(1), signed by participating state governors December 20, 2005, both available at [http://www.rggi.org/modelrule.htm].

[47] For more discussion regarding these issues, see CRS Report RL33799, *Climate Change: Design Approaches for a Greenhouse Gas Reduction Program*, by Larry Parker.

[48] 310 Massachusetts Code of Regulations 7.29.

[49] New Hampshire Clean Power Act (May 9, 2002), codified in New Hampshire Statute, Title X, Chapter 125-O (Multiple Pollutant Reduction Program).

[50] HB 3283, codified in Oregon Administrative Rules, Chapter 345, Division 24.

[51] Point Carbon, 2006, "Carbon Trading in the US: The Hibernating Giant," *Carbon Market Analyst*, September 13, 2006.

[52] HB 3141 (signed into law on March 31, 2004).

[53] Energy Information Administration, *Emissions of Greenhouse Gases in the United States 2005*, at [http://www.eia.doe.gov/].

[54] Based on 2001 data. Energy Information Administration, *Emissions of Greenhouse Gases in the United States 2004 (Table C2).*

[55] Only Vermont had a higher percentage from its transportation sector, but Vermont transportation emissions (3.7 million metric tons of carbon dioxide) are minor compared to California (210 million metric tons of carbon dioxide). Based on 2001 data. Energy Information Administration, *Emissions of Greenhouse Gases in the United States 2004 (Table C2).*

[56] See Clean Air Act Section § 209, codified at 42 U.S.C. § 7543.

[57] Clean Air Act § 177, codified at 42 U.S.C. § 7507.

[58] AB 1493 (or the California Vehicle Global Warming Law) was signed into law by Governor Gray Davis on July 22, 2002.

[59] Title 13, California Code of Regulations § 1961.1.

[60] In an EPA General Counsel memorandum (August 28, 2003), EPA took the position that the Clean Air Act does not authorize EPA to regulate greenhouse gases for the purpose of addressing climate change, and the agency has argued this position in court. See CRS Report RL32764, *Global Warming: The Litigation Heats Up*, by Robert Meltz.

[61] California Governor Schwarzenegger signed the legislation September 27, 2006.

[62] Earlier drafts of the legislation specifically cited the electric power, oil/gas, and cement industries, and landfills as significant emitters.

[63] The statute instructs CARB to regulate mobile sources if the 2004 mobile sources regulatory program (described above) does not remain in effect (presumably due to legal challenges).

[64] Defined as "every electrical corporation, electric service provider, or community choice aggregator serving end-use customers in the state." SB 1368 (codified in Public Utilities Code, Section 8340(h)).

[65] Defined as a "new ownership investment in baseload generation or a new or renewed contract with a term of five or more years, which includes procurement of baseload generation." SB 1368 (codified in Public Utilities Code, Section 8340(j)).

[66] SB 1368 was signed by the Governor on September 29, 2006.

[67] As technology advances, coal-fired plants might be able to reduce greenhouse gas emissions through carbon capture and sequestration (CCS). However, "there is relatively little experience in combining CO2 capture, transport and storage into a fully integrated CCS system. The utilization of CCS for large-scale power plants (the potential application of major interest) still remains to be implemented." Intergovernmental Panel on Climate Change (IPCC), 2005, *IPCC Special Report Carbon Dioxide Capture and Storage, Summary for Policymakers*, p. 8.

[68] California Energy Commission, 2006, *Inventory of California Greenhouse Gas Emissions and Sinks: 1990 to 2004*, Draft Staff Report, p. 12.

[69] See Holly, Chris, "California PUC Issues IOU Greenhouse Rules; Muni Nixes Coal Deal," *The Energy Daily*, December 15, 2006.

[70] California Public Utilities Commission, Decision 05-04-024, April 7, 2005.

[71] California Public Utilities Commission, Decision 04-12-048, December 16, 2004.

[72] Oregon and Colorado. See Pew Center on Global Climate Change website, at [http://www.pewclimate.org/states.cfm].

[73] Pew Center on Global Climate Change, "California PUC Carbon Adder" (case-study).

[74] This is based on 2001 data from the World Resources Institute, Climate Analysis Indicators Tool, at [http://cait.wri.org/].

[75] The RGGI states are Connecticut, Delaware, Maine, New Hampshire, New Jersey, New York, and Vermont. Maryland and Massachusetts are expected to join in 2007, and their emissions are included above.

[76] Stavins, Robert, 2006, "A Utility Safety Valve for Cutting CO2," *The Environmental Forum*, Volume 23, Number 2, March/April, 2006, p. 14.

[77] New York state is expected to be the first state to issue its rule implementing RGGI, according to statements made from state officials at a climate change workshop: Pew

Center on Climate Change, Innovative Approaches to Climate Change: A State and Regional Workshop, Washington, DC, October 10-11, 2006.

78 See Potts, Brian, 2006, "Regulating Greenhouse Gas Leakage: How California Can Evade the Impending Constitutional Attacks," *Electricity Journal*, Vol. 19, Issue 5, June 2006.

In: Global Climate Change Revisited
Editor: Horace M. Karling, pp. 47-59
ISBN 978-1-59454-039-4

Chapter 3

CLIMATE CHANGE: GREENHOUSE GAS REDUCTION BILLS IN THE 110TH CONGRESS*

Larry Parker

SUMMARY

A number of congressional proposals to advance programs that reduce greenhouse gases have been introduced in the 110th Congress. Proposals receiving particular attention would create market-based greenhouse gas reduction programs along the lines of the trading provisions of the current acid rain reduction program established by the 1990 Clean Air Act Amendments. This paper presents a side-by-side comparison of the major provisions of those bills and includes a glossary of common terms.

INTRODUCTION

Climate change is generally viewed as a global issue, but proposed responses generally require action at the national level. In 1992, the United States ratified the United Nations Framework Convention on Climate Change (UNFCCC), which called on industrialized countries to take the lead in reducing the six primary greenhouse gases to 1990 levels by the year 2000.[1] For more than a decade, a variety of voluntary and regulatory actions have been proposed or undertaken in the United States, including monitoring of power plant carbon dioxide emissions, improved appliance efficiency, and incentives for developing renewable energy sources. However, carbon dioxide emissions have continued to increase.

In 2001, President George W. Bush rejected the Kyoto Protocol, which called for legally binding commitments by developed countries to reduce their greenhouse gas emissions.[2] He also rejected the concept of mandatory emissions reductions. Since then, the Administration has focused U.S. climate change policy on voluntary initiatives to reduce the growth in

---

* Excerpted from CRS Report RL33846 dated January 31, 2007.

greenhouse gas emissions. In contrast, in 2005, the Senate passed a Sense of the Senate resolution on climate change declaring that a mandatory, market-based program to slow, stop, and reverse the growth of greenhouse gases should be enacted at a rate and in a manner that "will not significantly harm the United States economy" and "will encourage comparable action" by other nations.[3]

A number of congressional proposals to advance programs designed to reduce greenhouse gases have been introduced in the 110th Congress. These have generally followed one of three tracks. The first is to improve the monitoring of greenhouse gas emissions to provide a basis for research and development and for any potential future reduction scheme. The second is to enact a market-oriented greenhouse gas reduction program along the lines of the trading provisions of the current acid rain reduction program established by the 1990 Clean Air Act Amendments. The third is to enact energy and related programs that would have the added effect of reducing greenhouse gases; an example would be a requirement that electricity producers generate a portion of their electricity from renewable resources (a renewable portfolio standard). This report focuses on the second category of bills.

PROPOSED LEGISLATION IN 110TH CONGRESS

In the 110th Congress, four bills have been introduced that would impose controls on emissions of greenhouse gases. A comparison of major provisions is provided in Appendix 1.

S. 280, introduced by Senator Lieberman, would cap emissions of the six greenhouse gases specified in the United Nations Framework Convention on Climate Change, at reduced levels, from the electric generation, transportation, industrial, and commercial sectors — sectors that account for about 85% of U.S. greenhouse gas emissions. The reductions would be implemented in four phases, with an emissions cap in 2012 based on the affected facilities' 2004 emissions (for an entity that has a single unit that emits more than 10,000 metric tons of carbon dioxide equivalent); the cap steadily declines until it is equal to one-third of the facilities' 2004 levels. The program would be implemented through an expansive allowance trading program to maximize opportunities for cost-effective reductions, and credits obtained from increases in carbon sequestration, reductions from non-covered sources, and acquisition of allowances from foreign sources could be used to comply with 30% of reduction requirements. The bill also contains an extensive new infrastructure to encourage innovation and new technologies.

S. 309, introduced by Senator Sanders, would cap greenhouse gas emissions on an economy-wide basis beginning in 2010. Beginning in 2020, the country's emissions would be capped at their 1990 levels, and then proceed to decline steadily until they were reduced to 20% of their 1990 levels in the year 2050. The EPA has the discretion to employ a market-based allowance trading program or any combination of cost-effective emission reduction strategies. The bill also includes new mandatory greenhouse gas emission standards for vehicles and new powerplants, along with a new energy efficiency performance standard. The bill would establish a renewable portfolio standard (RPS) and a new low-carbon generation requirement and trading program.

S. 317, introduced by Senator Feinstein, would cap greenhouse gas emissions from electric generators over 25 megawatts. Beginning in 2011, affected generators would be capped at their 2006 levels, declining to 2001 levels by 2015. After that, the emission cap would decline 1% annually until 2020, when the rate of decline would increase to 1.5%. The allowance trading program includes an allocation scheme that provides for an increasing percentage of all allowances to be auctioned, with 100% auctioning in 2036 and thereafter. The cap-and-trade program allows some of an entity's reduction requirement to be meet with credits obtained from foreign sources and a variety of other activities specified in the bill.

H.R. 620, introduced by Representative Olver, is a substantially modified version of S. 280. Using the same basic structure as S. 280, the emission caps under H.R. 620 are more stringent. Reductions from affected sectors (electric generation, transportation, industrial, and commercial) would be set at 2004 levels in 2012 and then steadily decline until the cap is equal to about one-fourth of facilities' 2004 levels. Although H.R. 620 permits affected entities to comply with the reduction requirements with credits from foreign sources, sequestration, and reductions from non-covered entities, these sources are limited to 15% of the source's reduction requirement.

ENDNOTES

[1] Under the United Nations Framework Convention on Climate Change (UNFCCC), those gases are carbon dioxide (CO_2), methane (CH_4), nitrous oxide (N_2O), hydrofluorocarbons (HFCs), perfluorocarbons (PFCs), and sulfur hexafluoride (SF_6). Some greenhouse gases are controlled under the Montreal Protocol on Substances that Deplete the Ozone Layer, and are not covered under UNFCCC.

[2] For further information, see CRS Report RL30692, *Global Climate Change: The Kyoto Protocol*, by Susan R. Fletcher.

[3] S.Amdt. 866, passed by voice vote after a motion to table failed 43-54, June 22, 2005.

Appendix 1. Comparison of Key Provisions of Greenhouse Gas Reduction Bills

Topic	S. 280 (Lieberman)	S. 309 (Sanders)	S. 317 (Feinstein)	H.R. 620 (Olver)
Emission reduction/ limitation scheme	Absolute cap on total emissions from all covered entities in the electric power, transportation, industry, and commercial sectors.	Absolute cap on total emissions economy-wide.	Absolute cap on total emissions from covered electric generators.	Absolute cap on total emissions from all covered entities in the electric power, transportation, industry, and commercial sectors.
Specific emissions limits	Beginning in 2012, emissions from covered entities are capped at 6.13 billion metric tons, minus 2012 emissions from non-covered entities. Beginning in 2020, emission cap declines to 5.239 billion metric tons, minus 2020 emissions from non- covered entities. Beginning in 2030, emission cap declines to 4.1billion metric tons, minus 2030 emissions from non-covered entities. Beginning in 2050, emission cap further declines to 2.096 billion metric tons, minus annual emissions from non-covered entities.	Beginning in 2010, emissions economy-wide to be reduced 2% annually. Beginning in 2020, emission cap on economy-wide basis set at 1990 level, with declining emission caps of 26.7% below 1990 levels in 2030 and 53.3% in 2040. Beginning in 2050, emission cap set at 80% below 1990 levels.	Beginning in 2011, emissions from affected electric generators capped at 2006 levels. Beginning in 2015, emissions from affected electric generators capped at their 2001 levels, declining 1% annually until 2020. Beginning in 2020, emission cap declines 1.5% annually.	Beginning in 2012, emissions from covered entities are capped at 6.15 billion metric tons, minus 2012 emissions from non-covered entities. Beginning in 2020, emission cap declines to 5.232 billion metric tons, minus 2020 emissions from non-covered entities. Beginning in 2030, emission cap declines to 3.858 billion metric tons, minus 2030 emissions from non-covered entities. Beginning in 2050, emission cap further declines to 1.504 billion metric tons, minus annual emissions from non-covered entities.
Greenhouse gases defined	Carbon dioxide, methane, nitrous oxide (N_2O), hydrofluorocarbons (HFCs), perfluorocarbons (PFCs), and sulfur hexafluoride (SF_6).	Same six gases as S. 280.	Same six gases as S. 280.	Same six gases as S. 280.

Covered entities	In metric tons of carbon dioxide equivalents: any electric power, industrial, or commercial entity that emits over 10,000 metric tons carbon dioxide equivalent annually from any single facility owned by the entity; any refiner or importer of petroleum products for transportation use that, when combusted, will emit over 10,000 metric tons annually; and any importer or producer of HFCs, PFCs, or SF_6 that, when used, will emit over 10,000 metric tons of carbon dioxide equivalent.	EPA promulgates rule within two years of enactment that applies the most cost-effective reduction options on sources or sectors to achieve reduction goals.	Any fossil fuel-fired electric generating facility that has a capacity of greater than 25 megawatts and generates electricity for sale, including cogeneration and government-owned facilities.	In metric tons of carbon dioxide equivalents: any electric power, industrial, or commercial entity that emits over 10,000 metric tons carbon dioxide equivalent annually from any single facility owned by the entity; any refiner or importer of petroleum products for transportation use that, when combusted, will emit over 10,000 metric tons annually; and any importer or producer of HFCs, PFCs, or SF6 that, when used, will emit over 10,000 metric tons of carbon dioxide equivalent.
Responsible agency	Environmental Protection Agency (EPA).	Environmental Protection Agency (EPA).	Environmental Protection Agency (EPA).	Environmental Protection Agency (EPA).
General allocating and implementing strategy	A tradeable allowance system is established: EPA shall determine allocations based on several economic, equity, and sector-specific criteria, including economic efficiency, competitive effects, and impact on consumers. Allowances are to be allocated upstream to refiners and importers of transportation fuel, along with producers of HFCs, PFCs, and SF_6, and downstream to electric generation, industrial, and commercial entities. Allocations to covered entities are provided at no cost.	Tradeable allowance system permitted. In implementing reduction program, EPA shall select the most cost-effective emission reduction strategies. EPA shall allocate to various sectors and interests any allowances that are not allocated to affected entities, including households, dislocated workers, energy efficiency and renewable energy activities, sequestration activities, and ecosystem protection activities.	Tradeable allowance system is established. Allocations to existing sources based on historic electricity output, and includes allowance allocations for incremental nuclear capacity and renewable energy, along with sequestration and early action provisions. From 2011 on, an increasing percentage of all allowances are to be auctioned, with 100% of allowances auctioned in 2036 and thereafter.	A tradeable allowance system is established: EPA shall determine allocations based on several economic, equity, and sector- specific criteria, including economic efficiency, competitive effects, and impact on consumers. Allowances are to be allocated upstream to refiners and importers of transportation fuel, along with producers of HFCs, PFCs, and SF_6, and downstream to electric generation, industrial, and commercial entities. Allocations to covered entities are provided at no cost.

APPENDIX 1. (CONTINUED)

Topic	S. 280 (Lieberman)	S. 309 (Sanders)	S. 317 (Feinstein)	H.R. 620 (Olver)
Public sale/auction of allowances	EPA shall determine the number of allowances allocated to the Climate Change Credit Corporation (CCCC) (established by the bill). EPA shall allocate to the CCCC allowances before 2012 to auction to raise revenue for technology deployment and dissemination. The CCCC may buy and sell allowances, and use the proceeds to reduce costs borne by consumers and other purposes. (See "Revenue recycling" below.)	EPA may choose to provide for trustees to sell allowances for the benefit of entities eligible to receive assistance under the proposal (see above).	From 2011 on, an increasing percentage of all allowances are to be auctioned, with 100% of allowances auctioned in 2036 and thereafter. Revenues from the auction are to be deposited in the Climate Action Trust Fund created by the Department of the Treasury.	EPA shall determine the number of allowances allocated to the Climate Change Credit Corporation (CCCC) (established by the bill). The CCCC may buy and sell allowances, and use the proceeds to reduce costs borne by consumers and other purposes. (See "Revenue recycling" below.)
Cost-limiting safety valve	No explicit provision.	No explicit provision. However, if the President determines a national security emergency exists, the President may temporarily adjust, suspend, or waive any regulation promulgated under this program (subject to judicial review).	No explicit provision. However, limited borrowing against future reductions is permitted if EPA determines allowance prices have reached and sustained a level that is or will cause significant harm to the U.S. economy. Also, EPA may increase to 50% the share of international credits that can be used in such cases.	No explicit provision.
Other market trading system features	Up to 30% of required reductions may be achieved through credits obtained through pre-certified international emissions trading programs, approved reduction projects in developing countries, domestic carbon	Market trading systems incorporated into Renewable Portfolio Standard and new low-carbon generation requirement.	Up to 25% (50% for new affected units) of required reductions may be achieved with credits obtained through EPA-approved foreign government programs	Up to 15% of required reductions may be achieved through credits obtained through pre-certified international emissions trading programs, approved reduction projects in developing countries,

	sequestration, and reductions from non-covered entities. Borrowing against future reductions is permitted.		developed under United Nations Framework Convention on Climate Change (UNFCCC) protocols. Limited borrowing against future reductions is permitted if EPA determines allowance prices have reached and sustained a level that is causing or will cause significant harm to the U.S. economy. Also, EPA may increase to 50% the share of international credits that can be used in such cases.	domestic carbon sequestration, and reductions from non-covered entities. Borrowing against future reductions is permitted.
Banking	Banking of allowances is permitted; allowances may be saved for use in future years.	No specific prohibition on banking.	Banking of allowances is permitted; allowances may be saved for use in future years.	Banking of allowances is permitted; allowances may be saved for use in future years.
Early reduction credits and bonus credits	Entities with registered emission reductions achieved before 2012 may receive allowances for them, including reductions achieved under more stringent mandatory state programs. For the time period 2012-2017, entities that have entered into an agreement with EPA to reduce emissions to 1990 levels by 2012 are entitled to additional allowances to cover their additional reductions and are allowed to achieve 40% of their reduction requirement (as opposed to 30%; see above) through international emissions trading and projects, sequestration, or reductions by non-covered entities.	Reductions previously achieved under state programs that are at least as stringent as a federal trading program may be recognized by the federal program. Entities that demonstrate reductions achieved early (but not before 1992) that are as verifiable as reductions under a federal trading program may be recognized by the federal program.	Entities with reductions achieved from 2000 through 2010 shall receive credits under specific criteria, including EPA rules that ensure reductions are real, additional, verifiable, enforceable, and permanent, and that they were reported under either 1605(b) of the 1992 Energy Policy Act, or according to a state or regional registry. Quantity of credits given is limited to 10% of the 2011 allowance allocation.	Entities with registered emission reductions achieved before 2012 may receive allowances for them. For the time period 2012-2017, entities that have entered into an agreement with EPA to reduce emissions to 1990 levels by 2012 are entitled to additional allowances to cover their additional reductions and are allowed to achieve 35% of their reduction requirement (as opposed to 15%; see above) through international emissions trading and projects, sequestration, or reductions by non-covered entities.

APPENDIX 1. (CONTINUED)

Topic	S. 280 (Lieberman)	S. 309 (Sanders)	S. 317 (Feinstein)	H.R. 620 (Olver)
Revenue recycling	Revenues generated by allowance auctions and trading proceeds are received by a new Climate Change Credit Corporation (CCCC). Activities to be funded include mechanisms to reduce consumer costs and to assist dislocated workers, low-income persons, and affected communities, along with programs to encourage deployment of new technology and wildlife restoration. Allocations to the CCCC are to be determined by EPA based on the funding needs of the advanced technologies demonstration and deployment programs. Further, at least 50% of revenue received must be used for technology deployment.	Allowances may be allocated by EPA to households, dislocated workers, energy efficiency and renewable energy activities, sequestration activities, and ecosystem protection activities.	Revenues generated from the auction are to be deposited in the Climate Action Trust Fund created by the Department of the Treasury. Activities to be funded include an Innovative Low- and Zero-emitting Carbon Technologies Program, a Clean Coal Technologies Program, and an Energy Efficiency Technology Program, along with research and development. Adaption and mitigation activities to be funded include affected workers and communities, and fish and wildlife habitat.	Revenues generated by allowance auctions and trading proceeds are received by a new Climate Change Credit Corporation (CCCC). Activities to be funded include mechanisms to reduce consumer costs and to assist dislocated workers and affected communities, along with programs to encourage deployment of new technology and wildlife restoration.
Penalty for non-compliance	Excess emission penalties are equal to three times the market price for allowances on the last day of the year at issue.	Existing enforcement provisions of Section 113 of the Clean Air Act are extended to program.	$100 per excess ton indexed to inflation plus a 1.3 to 1 offset from future emissions allowances. If the market price for an allowance exceeds $60, the penalty is $200 per excess ton, adjusted for inflation.	Excess emission penalties are equal to three times the market price for allowances on the last day of the year at issue.
Other key provisions	Provisions include studies of research on abrupt climate change	Provisions include mandatory greenhouse gas	Establishes program to encourage offsets from the	Provisions include studies of the impact of climate change on

	and impact of climate change on the world's poor, among others, and creation of a national greenhouse gas database. A new Innovation Infrastructure is created, along with program initiatives to promote less carbon-intensive technology, adaption, sequestration, and related activities. Requires periodic review of target adequacy by the Under Secretary of Commerce for Oceans and Atmosphere.	emission standards for vehicles by 2010, for new electric powerplants that begin operation after December 31, 2011, and a new energy efficiency performance standard. Establishes a Renewable Portfolio Standard and credit program. Establishes a new low-carbon generation requirement and trading program. Requires periodic review of target adequacy by the National Academy of Sciences.	agricultural sector. Offset credits available for agricultural, forestry, grazing, and wetlands management, sequestration projects, or practices that meet specific criteria in the proposal. Offset credits also available for approved emission reduction offset projects from a variety of activities listed in the proposal. Requires periodic review of target adequacy by EPA taking into account the recommendations of a newly established Climate Science Advisory Panel.	coastal ecosystems and communities, and the world's poor, among others; assessment of adaptation technologies; and creation of a national greenhouse gas database. Requires periodic review of target adequacy by the Under Secretary of Commerce for Oceans and Atmosphere.

APPENDIX 2. COMMON TERMS

Allocation Schemes (Upstream and Downstream)

Regulatory approaches to allocating allowances (as opposed to auction schemes) can choose different points and participants along the production process to assign allowances and the resulting compliance responsibility. *Upstream allocation schemes* establish emission caps and assign allowances at a production, importation, or distribution point of products that will eventually produce greenhouse emissions further down the production process. For example, in the natural gas sector, emission caps could be established and allowances assigned at processing facilities where facilities and participants shrink from about 400,000 wells and 8,000 companies to 500 plants and 200 companies. In contrast, *downstream allocation schemes* establish emission caps and assign allowances at the point in the process where the emissions are emitted. In the case of the natural gas industry, to achieve the same coverage as the upstream scheme, this would involve assigning allowances to natural gas-fired electric generators, industry, and even residential users. Thus, some downstream proposals choose either to exempt certain sectors (such as residential use) from a cap-and-trade program or to employ a hybrid allocation scheme where some of the allowances are allocated upstream and others downstream (such as the electric generators).

Allowance

An allowance is generally defined as a limited authorization by the government to emit 1 ton of pollutant. In the case of greenhouse gases, an allowance generally refers to a metric ton of carbon dioxide equivalent. Although used generically, an *allowance* is technically different from a *credit*. A credit represents a ton of pollutant that an entity has reduced in excess of its legal requirement. However, the terms tend to be used interchangeably, along with others, such as *permits*.

Auctions

Auctions can be used in market-based pollution control schemes in several different ways. For example, Title IV of the 1990 Clean Air Act Amendments uses an annual auction to ensure the liquidity of the credit trading program. For this purpose, a small percentage of the credits permitted under the program are auctioned annually, with the proceeds returned to the entities that would have otherwise received them. Private parties are also allowed to participate. A second possibility is to use an auction to raise revenues for a related (or unrelated) program. For example, the Regional Greenhouse Gas Initiative (RGGI) is exploring an auction to implement its public benefit program to assist consumers or pursue strategic energy purposes. A third possibility is to use auctions as a means of allocating some, or all, of the credits mandated under a GHG control program. Obviously, the impact that an auction would have on cost would depend on how extensively it was used in any GHG control program, and to what purpose the revenues were expended.

Banking

Although allowances are generally allocated on an annual basis, most cap-and-trade programs do not require participants to either use the allowance that year or else lose it. Under many proposals, allowances can be banked by the receiving participant (or traded to another participant who can use or bank it) to be used or traded in a future year. Banking reduces the absolute cost of compliance by making annual emission caps flexible over time. The limited ability to shift the reduction requirement across time allows affected entities to better accommodate corporate planning for capital turnover, allow for technological progress, control equipment construction schedules, and respond to transient events such as weather and economic shocks.

Bubble

A bubble is a regulatory device that permits two or more sources of pollutants to be treated as one for the purposes of emission compliance.

Cap-and-trade Program

A cap-and-trade program is based on two premises. First, a set amount of pollutant emitted by human activities can be assimilated by the ecological system without undue harm. Thus, the goal of the cap-and-trade program is to impose a ceiling (i.e., an emissions cap) on the total emissions of that pollutant at a level below the assimilative capacity. Second, a market in pollution licenses (i.e., allowances) between polluters is the most cost-effective means of reducing emissions to the level of the cap. This market in allowances is designed so that owners of allowances can trade those allowances with other emitters who need them or retain (bank) them for future use or sale. In the case of the sulfur dioxide program contained in the 1990 Clean Air Act Amendments, most allowances were allocated free by the federal government to utilities according to statutory formulas related to a given facility's historic fuel use and emissions; other allowances have been reserved by the government for periodic auctions to ensure market liquidity.

Carbon Tax

A carbon tax is generally conceived as a levy on natural gas, petroleum, and coal according to their carbon content, in the approximate ratio of 0.6 to 0.8 to 1, respectively. However, proposals have been made to impose the tax downstream of the production process when the carbon dioxide is actually released to the atmosphere. In contrast to a cap-and-trade program, in which the quantity of emissions is limited and the price is determined by an allowance marketplace, with a carbon tax, the price is limited and the quantity of emissions is determined by the participants based on the cost of control versus the cost of the tax.

Coverage

Coverage is the breadth of economic sectors covered by a particular greenhouse gas reduction program.

Emissions Cap

A mandated limit on how much pollutant (or greenhouse gases) an affected entity can release to the atmosphere. Caps can be either an *absolute cap*, where the amount is specified in terms of tons of emissions on an annual basis, or a *rate-based cap*, where the amount of emissions produced per unit of output (such as electricity) is specified but not the absolute amount released. Caps may be imposed on an entity, sector, or economy-wide basis.

Generation Performance Standard (GPS)

Also called an *output-based allocation*, allowances are allocated gratis to entities in proportion to their relative share of total electricity generation in a recent year.

Grandfathering

Grandfathering generally refers an allocation scheme in which allowances are distributed to affected entities on the basis of historic emissions. These allowances are generally distributed free-of-charge by the government to the affected entities. Grandfathering can also refer to entities that because of age or because they have met an earlier standard, or other factors, are exempted from a new regulatory requirement.

Greenhouse Gases

The six gases recognized under the United Nations Framework Convention on Climate Change are carbon dioxide (CO_2), methane (CH_4) nitrous oxide (N_2O), sulfur hexafluoride (SF_6), hydrofluorocarbons (HFC), and perfluorocarbons (PFC).

"No Regrets" Policy

A "no regrets" policy is one of establishing programs for other purposes that would have concomitant greenhouse gas reductions. Therefore, only those policies that reduce greenhouse gas emissions at no cost are considered.

Offsets

Offsets generally refer to emission credits achieved by activities not directly related to the emissions of an affected source. Examples of offsets would include forestry and agricultural activities that absorb carbon dioxide, and reduction achieved by entities that are not regulated by a greenhouse gas reduction program.

Revenue Recycling

Some greenhouse gas reduction programs create revenues through auctions, compliance penalties, or imposition of a carbon tax. Revenue recycling refers to how a program disposes of those revenues. How a program handles revenues received can have a significant effect on the overall cost of the program to the economy.

Safety Valve

Devices designed to prevent or to respond to unacceptably high compliance costs for greenhouse gas reductions. Generally triggered by prices in the allowance markets, safety valve approaches can include (1) a set price alternative to making reductions or buying allowances at the market price, (2) a slowdown in tightening the emissions cap, and (3) lengthening of the time allowed for compliance. Depending on the interplay between the emissions cap and safety valve and actual compliance costs, a safety valve can affect the integrity of the emissions cap.

Sequestration

Sequestration is the process of capturing carbon dioxide from emission streams or from the atmosphere and then storing it in such a way as to prevent its release to the atmosphere.

In: Global Climate Change Revisited
Editor: Horace M. Karling, pp. 61-82
ISBN 978-1-59454-039-4

Chapter 4

CLIMATE CHANGE: THE EUROPEAN UNION'S EMISSIONS TRADING SYSTEM (EU-ETS)*

Larry Parker

SUMMARY

The European Union's (EU's) Emissions Trading System (ETS) is a cornerstone of the EU's efforts to meet its obligation under the Kyoto Protocol. It covers more than 11,500 energy intensive facilities across the 25 EU member countries, including oil refineries, powerplants over 20 megawatts (MW) in capacity, coke ovens, and iron and steel plants, along with cement, glass, lime, brick, ceramics, and pulp and paper installations. Covered entities emit about 45% of the EU's carbon dioxide emissions. The trading program does not cover emissions of non-CO2 greenhouse gases, which account for about 20% of the EU's total greenhouse gas emissions. The first trading period began January 1, 2005. A second trading period is scheduled to begin in 2008, with a third one planned for 2013. In deciding on its trading program, the European Commission (EC) adopted a "learning-by-doing" approach to prepare the EU for the Kyoto Protocol's emission limitations. The EU does not have major experience with emissions trading, and the EC felt that an initial program beginning in 2005 would give the EU practical familiarity in operating such a system.

At first glance, it would appear that the EU may have little difficulty meeting its Kyoto Protocol requirements during the second trading period. The anticipated deficit between the second trading period for the original 15 Member States can be covered by trading with the 10 newer Member States that anticipate a surplus. Also, credits are likely to be available through Joint Implementation (JI) and Clean Development Mechanism (CDM) projects sanctioned under the Protocol.

However, there are other considerations. The availability of surplus credits created via JI and CDM is restricted by the EC requirement that such credits be "supplemental" to a country's domestic efforts. Each country is to spell out what "supplemental" means in its

* Excerpted from CRS Report RL33581 dated July 31, 2006.

National Allocation Plans (NAPs) for the second trading period. Individual countries are likely to define that term differently — restricting allowance trades and purchases in some countries.

Another consideration is the overall commitment of the Kyoto Protocol. As noted earlier, the ETS covers only a percentage of the overall greenhouse gas emissions in the various Member States of the EU. Some sectors not covered by the ETS may grow faster than sectors covered by it, creating difficulties for compliance. In particular, the transportation area is already a source of concern.

A final consideration for the ETS is its suitability for directing long-term investment toward a low-carbon future — the ultimate goal of any climate change program. It is too early to tell whether the ETS's market signal and individual countries' NAPs will move investment in the appropriate direction. The early signs are not particularly encouraging, with the 2005-2008 NAPs producing an over-allocation of allowances and one major Member State (Germany) attempting to direct its second NAP toward carbon-intensive, coal-fired electric-generating facilities rather than low-carbon alternatives. Reluctance by countries to redirect their NAPs and an inconsistent price signal from the ETS make the long-term effect of the ETS uncertain.

OVERVIEW

Climate change is generally viewed as a global issue, but proposed responses typically require action at the national level. With the 1997 Kyoto Protocol now in force, countries that ratified the protocol are developing appropriate implementation strategies to begin reducing their emissions of greenhouse gases.1 In particular, the European Union (EU) has decided to use an emissions trading scheme (called a "cap-and-trade" program), along with other market-oriented mechanisms permitted under the Protocol, to help it achieve compliance at least cost.2 The decision to use emission trading to implement the Kyoto Protocol is at least partly based on the successful emissions trading program used by the United States to implement its sulfur dioxide (acid rain) control program contained in Title IV of the 1990 Clean Act Amendments.3

The EU's Emissions Trading System (ETS) is a cornerstone of the EU's efforts to meet its obligation under the Kyoto Protocol. It covers more than 11,500 energy intensive facilities across the 25 EU Member countries, including oil refineries, powerplants over 20 megawatts (MW) in capacity, coke ovens, and iron and steel plants, along with cement, glass, lime, brick, ceramics, and pulp and paper installations. Covered entities emit about 45% of the EU's carbon dioxide emissions. The trading program does not cover emissions of non-CO2 greenhouse gases, which account for about 20% of the EU's total greenhouse gas emissions. The first trading period began January 1, 2005. A second trading period is scheduled to begin in 2008, covering the period of the Kyoto Protocol, with a third one planned for 2013.4

Under the Kyoto Protocol, the then-existing 15 nations of the EU agreed to reduce their emissions by 8% from 1990 levels under a collective arrangement called a "bubble." By 2001, collective greenhouse gas emissions in the EU were 2.3% below 1990 levels, mostly the result of a structural shift from coal to natural gas in the United Kingdom and the incorporation of East Germany into West Germany. Several countries, including Ireland,

Spain, and Portugal, experienced emissions growth of over 30% during this period.5 In light of the Kyoto Protocol targets, the EU adopted a directive establishing the EU-ETS that entered into force October 13, 2003.6 The importance of emissions trading was elevated by the accession of 10 additional central and eastern Europe countries to EU membership in May 2004. Collectively, these 10 countries' greenhouse gas emissions dropped 22.6% from 1990-2001, with only Slovenia's emissions increasing during that time (10.4%). This expansion of the EU trading zone to 25 countries greatly increases the opportunities for cost-effective allowance trades.

In deciding on its trading program, the European Commission (EC) adopted a "learning-by-doing" approach to prepare the EU for the Kyoto Protocol's emissions limitations. The EU does not have major experience with emissions trading, and the EC felt an initial program beginning in 2005 would give the EU practical familiarity in operating such a system. The EC also wanted the most comprehensive program possible. As stated in its "Green Paper":

The wider the scope of the system, the greater will be the variation in the costs of compliance of individual companies, and the greater the potential for lowering costs overall. This argues in favour of a comprehensive trading scheme across different Member States covering all 6 greenhouse gases and sinks, and encompassing all emissions sources.7

Economic analysis conducted by the European Commission confirms the potential cost-saving available from a comprehensive trading scheme. As shown in Table 1, a comprehensive trading program is estimated by the EC to reduce Kyoto compliance costs to EU countries by 3 billion euro, or one-third over a compliance scenario that does not include trading among Member countries, and by 0.9 billion euro, or 13% below the estimate cost of compliance with the trading scheme ultimately chosen by the EC.

Table 1. Cost of Reaching Kyoto Target to EU Member States in 2010 (in billions of 1999 euro)

No Trading Among EU Member States	EU-wide Trading Among Energy Producers	EU-wide Trading Among Energy Producers and Energy-Intensive Industries	EU-wide Trading Among All Sectors
9.0	7.2	6.9	6.0

Source: EC *Green Paper* (Mar. 8, 2000), p. 27.

For a variety of reasons, the EC chose a trading system with limited coverage rather than a comprehensive system covering all sources and gases. Some European analysts have noted that EU politics played an important role in preventing serious consideration of a comprehensive program. As noted by Boemare and Quirion:

> A significantly wider coverage could have been provided only by an upstream system, which had been excluded by the [European] Commission at the beginning of the process. The reason was again political: an upstream scheme would have too much looked like a tax.[8]

Not surprisingly, this reason was not employed by the EC in explaining its decision to create a less comprehensive trading scheme at this time. As stated by the EC:

> ... there are sound scientific and practical reasons why the Community might not wish to establish a comprehensive scheme at this stage. There are considerable uncertainties surrounding the emissions of the fluorinated gases [HFC, PFC, SF_6] and the absorption of carbon dioxide by sinks. Allocating allowances, monitoring emissions and enforcing compliance of small mobile emitters, such as private cars, raise complex technical and administrative issues.[9]

For determining the size of the trading program, the EC looked at five criteria: (1) environmental effectiveness, (2) economic efficiency, (3) the potential effects on competition, (4) feasibility, and (5) existence of alternative policies and measures. It felt that starting with a relatively small number of economic sectors and sources that contribute significantly to total emissions and for which trading could reduce cost significantly would "substantially" satisfy these criteria.[10] As noted, the six sectors chosen emit about 45% of the EU15's CO_2 emissions (which are about 80% of the EU's total greenhouse gas emissions). The coverage for individual countries varies widely; only 20% of France's greenhouse gas emissions are covered, compared with 69% of Estonia's emissions.[11]

Implementing the ETS: National Allocation Plans

National Allocation Plans (NAPs) are central to the EU's effort to achieve its Kyoto obligations. Each Member of the EU must submit a NAP that lays out its allocation scheme under the ETS, including individual allocations to each affected unit. For the first trading period, each country had to prepare a NAP by March 31, 2004 (May 1, 2004 for the 10 new EU Members). NAPs for the second period were due June 30, 2006. These NAPs are assessed by the EC to determine compliance with 11 criteria (12 for the second period) delineated in an annex to the emissions trading directive.[12] Criteria include requirements that the emissions caps and other measures proposed by the state are sufficient to put it on the path toward its Kyoto target, protections against discrimination between companies and sectors, along with provisions for new entrants, clean technology, and early reduction credits. For the second period, the NAP must guarantee Kyoto compliance.

For the first period, the EC approved most of the necessary NAPs by the end of 2004. The last NAP was approved June 20, 2006 (from Greece). In general, the primary problem the EC found with NAPs that resulted in revisions were excessive allocation of allowances and state efforts to permit "ex-post adjustments" to their allocations. Excessive allocation problems resulted from states that left a gap in how they would achieve their target, to be filled with measures to be defined later; insufficiently delineated plans to purchase allowances; and unrealistic economic or emissions growth assumptions. Ex-post adjustments by states are not allowed; such adjustments are seen by the EC as potentially disruptive to the emissions market and creating uncertainty for companies.

RESULTS FROM THE FIRST YEAR

Emissions Levels

For the 2005-2007 period, the European Union is not attempting to meet the Kyoto Protocol but to get experience with emissions trading with some modest emissions targets (i.e., to put the EU on the path toward meeting the Kyoto requirements). Table 2 provides the national emissions allocations and 2005 emissions levels for 21 EU countries as recorded by the Community Independent Transaction Log (CITL) by the compliance deadline of April 30, 2006.[13] For the first trading period, the 21 countries have allocated an annual average of 1.8295 billion allowances and set aside 73.4 million allowances for allocation to new sources or for auctions. Verified emissions in 2005 for covered sources is 1.7853 billion metric tons, according to the CITL.[14] The 44.2 million allowances allocated in excess of actual 2005 emissions have been characterized by EU's Environment Commissioner as an "over-allocation" of allowances[15] and is considered responsible for a significant drop in allowances prices in May, 2006. The 2005 emissions total reflects emissions from 8,980 sources representing more than 99% of the allowances allocated. As of April 30, 849 sources in the 21 countries had not surrendered sufficient allowances. The EC will determine whether the insufficiency is the result of technical difficulties in national registries, tardiness, or noncompliance. Noncomplying sources are subject to a 40-euro penalty for each ton of emissions in excess of surrendered allowances under the ETS.

Some commentators have suggested that annual average 2005-2007 allocations that are actually 44.1 million metric tons higher than the reported 2005 emissions are neither putting the EU on the path to the Kyoto requirements nor developing the trading market. In addition, this "over-allocation" does not include the 73.4 million metric tons of allowances held in reserve by the various countries for new entrants. The Climate Action Network (CAN), a network of 365 nongovernmental organizations, stated the following:

> Emissions limits set by Member States for the first phase were a major disappointment. To ensure maximum environmental benefit of the ETS and the overall success of the system as a whole, they need to be strengthened considerably. The Kyoto targets require ambitious caps with absolute reductions for the phase 2008-2012.[16]

In general, the EC has seen the over-allocation issue as part of the "learning-by-doing" process that should help the EU in implementing the second trading period beginning in 2008. As stated in its press release:

> The new 2005 emissions data gives independently assessed installation-level figures for the first time and so provides Member States with an excellent factual basis for deciding upon the caps in their forthcoming national allocations plans for the second trading period, when the Kyoto targets have to be met. The plans are subject to approval by the Commission, which will also be making extensive use of the 2005 emissions data.[17]

Table 2. Summary Information Per Member State

Member State	CO_2 emissions for 2005 in tonnes	Installations that have not reported by 30 April	Share of installations with verified emissions reports	Installations covered[a]	Installations not in compliance on 30 April 2006	Annual average allocation in 2005 to 2007 in tonnes[b]	Annual average allocation not allocated at the outset in tonnes[c]
Austria	33,372,841	0	100.0%	199	0	32,674,905	330,050
Belgium	55,354,096	2	99.9%	309	2	59,853,575	2,545,876
Czech Republic[d]	82,453,727	39	98.4%	389		96,907,832	348,020
Denmark	26,090,910	2	98.9%	380	4	31,039,618	2,460,382
Estonia	12,621,824	0	100.0%	43	1	18,763,471	189,529
Finland	33,072,638	10	100.0%	578	19	44,587,032	862,952
France[d]	131,147,905	17	99.7%	1075		150,500,685	4,871,317
Germany	473,715,872	13	99.8%	1842	90	495,073,574	3,926,426
Greece	71,033,294	28	99.5%	141	29	71,135,034	3,286,839
Hungary	25,714,574	13	99.0%	229	19	30,236,166	1,424,738
Ireland	22,397,678	0	100.0%	109	0	19,238,190	3,081,180
Italy	215,415,641	208	95.4%	943	647	207,518,860	15,551,575
Latvia	2,854,424	1	99.9%	92	1	4,054,431	505,760
Lithuania	6,603,869	2	99.9%	93	4	11,468,181	797,213
Netherlands	80,351,292	0	100.0%	209	0	86,439,031	2,503,305
Portugal	36,413,004	1	99.9%	243	2	36,898,516	1,262,898
Slovak Republic[d]	25,237,739	0	100.0%	175		30,364,848	7,180
Slovenia	8,720,550	0	100.0%	98	0	8,691,990	66,667
Spain[d]	181,063,141		99.1%	800		162,111,391	13,162,130
Sweden	19,306,761	29	99.4%	705	31	22,530,831	678,149

Table 2. Summary Information Per Member State (Continued)

Member State	CO_2 emissions for 2005 in tonnes	Installations that have not reported by 30 April	Share of installations with verified emissions reports	Installations covered[a]	Installations not in compliance on 30 April 2006	Annual average allocation in 2005 to 2007 in tonnes[b]	Annual average allocation not allocated at the outset in tonnes[c]
United Kingdom	242,396,039	15	99.9%	768	16	209,387,854	15,527,484
Total	*1,785,337,819*		*99.1%*	*9,420*		*1,829,476,015*	*73,389,670*

Note: As all data are held in the CITL and national registries, no data are available for those Member States without an active registry.

a. The figures in this column indicate the number of installations with active registry accounts on 30 April 2006. They differ from figures communicated in earlier press releases because they are updated for installations that opted-out for the first trading period, opted-in, and installations without open accounts.

b. The figures in this column are allowances allocated to existing installations at the start of the scheme.

c. The figures in this column are allowances not allocated to existing installations at the start of the scheme but put aside mainly for new entrants and auctioning (in the cases of Denmark, Hungary, Ireland, and Lithuania).

d. Due to technical problems in the national registries of the Czech Republic, France, the Slovak Republic, and Spain, the CITL did not receive wholly reliable information on the installation level surrenders from these Member States. Therefore, some fields are empty for these Member States. All data represented in the table were communicated directly to the European Commission by the respective authorities of these Member States.

Market Activity, Prices, and Impact

According to Point Carbon's proprietary databases, the EU-ETS traded 362 million metric tons of CO2 in 2005, valued at 7.218 billion euro. Brokers were responsible for 57% of the volume, exchange markets did 15%, and bilateral transactions accounted for 28%. Of the exchange market volume, the European Climate Exchange (ECX) had the largest share at 63%, followed by Nord Pool with 24%, and Powernext with 7.9%.18

The average price for an allowance traded in 2005 was 19.9 euro, with brokered and exchanged allowances averaging 20.6 euro and bilateral transactions averaging 18.2 euro. However, allowance prices have been quite volatile since trading begin in 2005, as indicated in Figure 1 below.[19] In particular, allowance prices dropped from almost 30 euro to about 9-11 euro in April and May, sparked by a series of reported over-allocation of allowances in several Member States. By July 2006, allowance prices had recovered to about 17 euro in July.

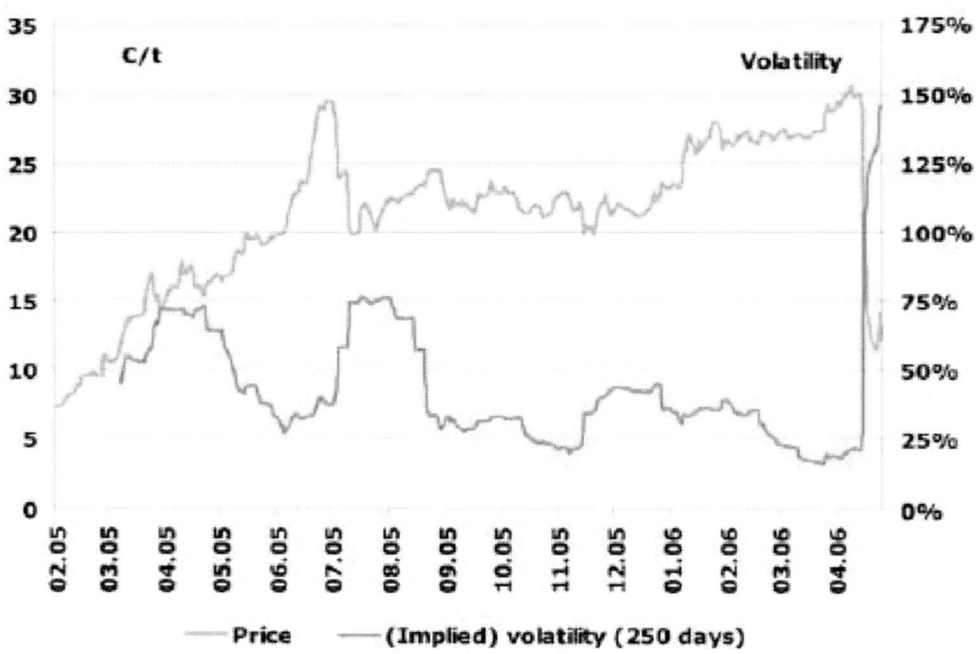

Note: EUA 2006, Vaws on 30 Day rolling price change. 9 May, 2006.
Source: Margus Kaasik, Eesti Energia, *Carbon Market: EUETS* (May 9, 2006), p. 14.

Figure 1. CO_2 Market: Even If No "Big" News Highly Volatile.

There are several reasons for the overall volatility in the allowance market. The EU-ETS is a maturing but still narrow market. Monthly volumes are increasing, but have never exceeded a 1.6% share of phase 1 allocations.20 Modest volume for a new system is not surprising; trading volumes under the U.S. Clean Air Act Title IV sulfur dioxide trading program were very thin in the beginning. Even after several years of operation, SO2 allowances prices can change unpredictably and inexplicably.21

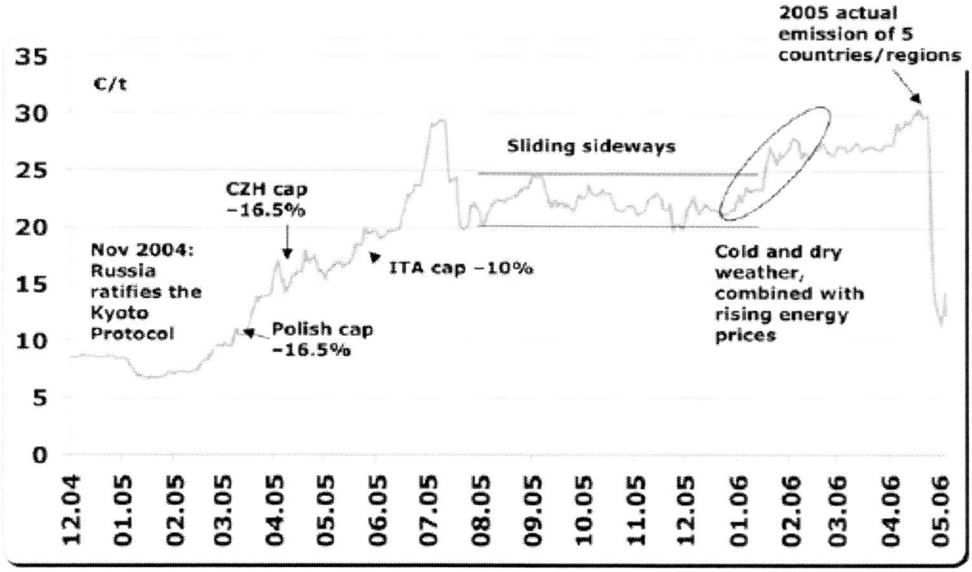

9 May, 2006

Source: Margus Kaasik, Eesti Energia, *Carbon Market: EUETS* (May 9, 2006), p. 14.

Figure 2. CO_2 Market: Large Price Changes in Very Short Amount of Time.

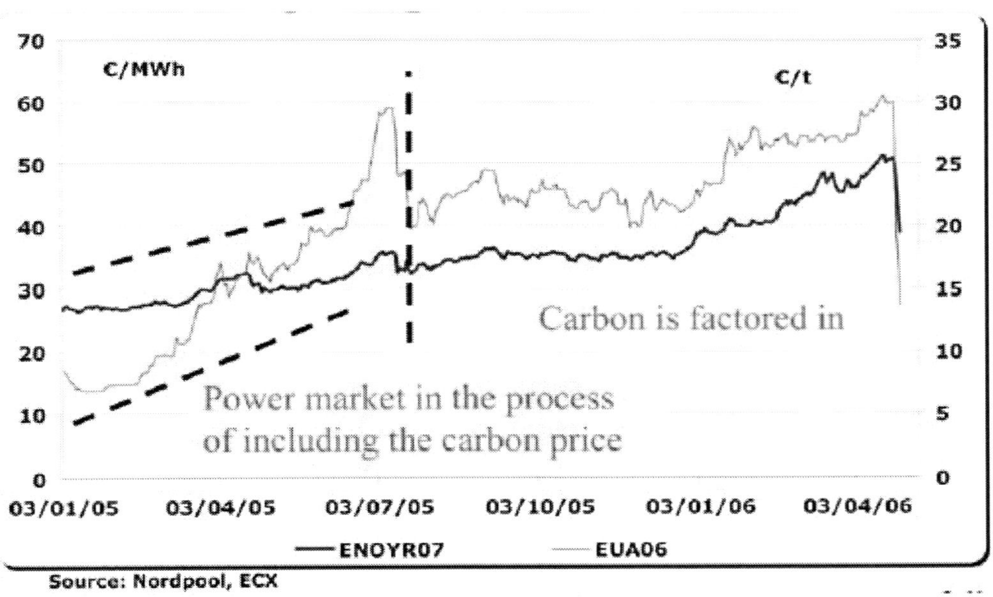

9 May, 2006

Source: Margus Kaasik, Eesti Energia, *Carbon Market: EUETS* (May 9, 2006) p. 11.

Figure 3. There is a Very Long-Term Correlation Between CO_2 and Electricity Price: Link Via Marginal Producer.

Some reasons for ETS allowance price derivatives are explicable. As illustrated by Figure 2, the ETS market responded to a variety of regulatory, climatic, and economic events over the first trading period. Regulatory events include the late approval of NAPs for several countries, along with the resulting over-allocation causing the sudden market correction in May 2006. Climatic events influencing prices include cold weather, which increased energy usage, and dry conditions, which decreased the availability of hydroelectric power.22

The primary economic influence on the ETS revolved around fuel prices. During 2005, Point Carbon analysis indicates 79% of the variance (R2) in allowance prices was explained by changes in fuel prices (particularly for electric power), with 23% of the variance explained by the weather. This linkage between allowance prices and the power market is not surprising, as the power sector conducted the majority of trades in 2005 and therefore significantly influenced price development.

Kaasik argues that the evidence from the past year indicates that fuel prices influence carbon prices, but not the reverse. Specifically, Kaasik sees allowance prices as a derivative of natural gas and coal pricing, at least in the short-term. Assuming natural gas-fired and coal-fired generation are the marginal cost suppliers of power, allowance prices will respond positively to increasing natural gas prices or decreasing coal prices. Likewise, allowance prices will respond negatively to decreasing natural gas prices or increasing coal prices. This creates a correlation over time between allowance prices and electricity price by influencing the marginal price of electricity. How this has evolved during the first trading period is illustrated by Figure 3.

Use of Clean Development Mechanism (CDM) and Joint Implementation (JI)

The EU-ETS has provisions for linking its trading scheme to the Joint Implementation (JI) and Clean Development Mechanism (CDM) components of the Kyoto Protocol for countries that have ratified it. These project-based instruments involve Annex 1 countries in the case of JI, and between Annex 1 and developing countries in the case of CDM.[23] The EC's linking directive allows operators to fulfil their allowance obligations under the EU-ETS using credits derived from JI and CDM projects. Their credits are equivalent to allowances in environmental and economic terms, but are not interchangeable. "Certified Emissions Reductions" (CERs) under the CDM must be issued by the Clean Development Mechanism Executive Board and may be used in either the first or banked for use in the second trading period.[24] Emissions Reduction Units (ERUs) under JI are transferred from one country to another — an exchange that cannot begin until the second trading period. Neither CERs nor ERUs are converted into EU-ETS allowances; rather, they are entered directly into the surrendered allowance table. There are other restrictions on the use of CERs and ERUs. In particular, for the second trading period the amount of CERs and ERUs that can used by an affected unit is limited by a percentage specified by its country. Several EU countries have established carbon funds to pursue JI and CDM opportunities.[25]

In general, CER and ERU credits have sold at a discount to ETS allowance prices. The degree of discount has depended on the riskiness of the project. CER and ERU credits are available only when the projects are completed. Thus, where buyers take the risk of non-delivery, such as an emissions reduction purchase agreement (ERPA), prices are in the range of 8-12 euro. In contrast, for CERs already issued, or where the sellers take the risk, prices are in the range of 13-15 euro.[26] The real impact of CDM and JI on the EU-ETS system will not be fully known until the second trading period, when EU demand for credits will increase substantially and other non-EU countries would be implementing their own Kyoto compliance strategies.

ISSUES

To assist its review of the ETS, the EC has surveyed stakeholders' viewpoints on ETS implementation and long-term issues.[27] The report surveys the viewpoints of participating companies, governments, industry associations, market intermediaries, and non-governmental organizations (NGOs) from June to September 2005. Asked which of 12 topics surrounding ETS implementation entities felt were most important to them, companies, industry associations, and governments all ranked topics such as emissions reduction targets, allocation rules, and rules for new entrants and closures as the most important — topics that all relate to long-term uncertainty. Table 3 indicates the top five topics according to governments surveyed, along with their corresponding ranking by other stakeholders. The five issues are discussed below.

Tightening of Emissions Caps

With the over-allocation issue in the first trading period, it is likely that the EC will take a harder stance in reviewing NAPs for the second trading period. The relationship between the ETS cap for the first trading period and the estimated ETS cap for the second trading period (the Kyoto Protocol requirements) varies substantially between countries, as illustrated in Table 4. In general, the original EU-15 countries have to reduce their emissions caps the most to meet their share of the EU's requirements under the Kyoto Protocol, with several countries facing double-digit percentage reductions. As indicated, EU-15 states, on average, have to reduce their emissions caps 6.8% (119 million metric tons) from their current levels to meet their requirements under the Kyoto Protocol-based second trading period. In contrast, as a group, the newer countries and the EU as a whole are in substantially better shape.[28]

Table 3. Importance of EU ETS Topics(ranking)

Topic	Govern-ments	Companies	Industry Associations	Market Intermediaries	NGOs
Emissions reduction targets	1st	1st	1st	3rd	1st
Further harmonization of allocation plans	2nd	3rd	2nd	Tied for 5th	Tied for 6th
Treatment of new entrants/ closures	3rd	2nd	5th	7th	Tied for 6th
Definition of combustion installations	4th	10th	Tied for 8th	Tied for 8th	Tied for 11th
Inclusion of sectors and gases	5th	Tied for 6th	Tied for 8th	Tied for 4th	3rd

Source: European Commission, *Review of EU Emissions Trading Scheme* (November 2005), p. 13.

Table 4. Comparison of 1st and 2nd Trading Period ETS Caps (in millions of metric tons of CO_2 Equivalent unless otherwise noted)

	Average Annual ETS Cap (first trading period)[a]	Estimated Annual ETS Cap (Kyoto Protocol)	Percentage Difference
Austria	33.0	24.6	-25.5%
Belgium	62.9	57.9	-8.0%
Czech Republic	97.6	118.6	21.6%
Denmark	33.5	24.9	-25.6%
Estonia	19.0	35.4	86.5%
Finland	45.5	37.5	-17.7%
France	156.5	159.6	2.0%
Germany	499.0	483.2	-3.2%
Greece	74.4	75.5	1.5%
Hungary	31.3	43.0	37.3%
Ireland	22.3	20.1	-9.7%
Italy	232.5	194.7	-16.3%
Latvia	4.6	10.1	119.8%
Lithuania	12.3	33.4	171.5%
Luxembourg	3.4	2.7	-19.4%
Netherlands	95.3	88.9	-6.7%
Poland	239.1	331.0	38.4%
Portugal	38.2	35.4	-7.2%
Slovakia	30.5	38.9	27.7%
Slovenia	8.8	8.3	-5.4%
Spain	174.4	142.8	-18.1%
Sweden	22.9	24.4	6.7%
UK	245.3	247.8	1.0%
EU-ETS 15	1,739.1	1,620.1	-6.8%
EU-ETS 23	2,182.3	2,239.0	2.6%

Source: Based on data provided in Annex 1, European Commission, "Further Guidance on Allocation Plans for the 2008 to 2012 trading period of the EU Emissions Trading Scheme" (Brussels, Dec. 12, 2005) p. 11.

a. These figures do not account for changes to the number of installations subsequent to the respective Commission decision (e.g., opt-ins or opt-outs of installations).

At first glance, it would appear that the EU would have little difficulty meeting its Kyoto Protocol requirements during the second trading period. The anticipated deficit between the second trading period for the original 15 Member States can be covered by trading with the newer Member States that anticipate a surplus. However, there are other considerations. First, countries with potential surpluses may want to retain at least some of that surplus to help fuel their countries' economic growth, possibly at the expense of a Member State that needs allowances. Second, the extent to which surplus credits would be created via JI, the EC linking directive, requires that such credits (including from CDM) be "supplemental" to a country's domestic efforts. Each country is to spell out what "supplemental" means in its NAP for the second trading period.

A third consideration is the overall commitment of the Kyoto Protocol. As noted earlier, the ETS only covers a percentage of the overall greenhouse gas emissions in the various Member States of the EU. The analysis provided in Table 3 assumes that the ETS will have to provide a proportional amount of that reduction based on 2003 emissions. However, some sectors not covered by the ETS may grow faster than sectors covered by the ETS, creating difficulties for compliance. In particular, the transportation area has become a major source of concern. The transportation sector is not a part of the ETS and is not likely to be included before the third trading period.29 Instead, transport controls are based on voluntary agreements with automobile manufacturers to improve fuel economy, fuel-economy labelling of cars, and promoting fuel efficiency by fiscal measures.30 The cornerstone is the agreements with automobile manufacturers to achieve improve new car fleet average CO_2 emissions rates. As announced in 1996, the objective of the EU Council of Environmental Ministers and European Parliament was to achieve a new car fleet average CO_2 emissions rate of 120 grams per kilometer (g CO_2/km) by 2005, or by 2010 at the latest.31

This objective was not met in 2005 and is unlikely to be met by 2010. Voluntary commitments by the European, Japanese, and Korean Automobile Manufacturers Associations in 1998 and subsequently endorsed by the EC set targets of 140 g CO_2/km by 2008/2009. At the end of 2005, the average new car emissions rate is about 160 g CO_2/km. The rate of reduction in CO_2/km for new cars would have to double for the automobile manufacturers to achieve their commitments, which appears unlikely.32 Indeed, despite improvements in emissions rates, CO_2 emissions in the EU continue to rise because of increased miles driven, increased size and weight of cars, and falling car occupancy rates. The EC 2005 progress report also notes that despite EU's effort to increase information on fuel efficiency and CO_2 emissions for consumers, the effectiveness of the effort seems low: "a significant impact on consumer's decisions could not yet be noticed."33

Attempts to balance the burden between ETS and non-ETS sectors have already created tension in Germany with respect to the second trading period. In its draft NAP for the second trading period (NAP II) submitted June 28, 2006, Germany proposes to reduce its ETS allocation from about 499 million metric tons to about 471 million metric tons, a decrease of 5.6%.34 However, Germany's 2005 emissions were only 474 million metric tons; thus the reduction is only 0.6% from last year's emissions. In addition, Germany is proposing to permit every new power station built between 2008 and 2012 to opt out from CO_2 caps and the ETS for 14 years, generally to encourage construction of coal-fired facilities. These emissions increases will have to be covered by the non-ETS sectors — commercial, residential, and transportation. An assessment done for Greenpeace International states that the German NAP II places "a disproportionate burden for emissions reductions on the non-ETS sectors" in terms of meeting its commitments under the EU Burden Sharing agreement.35 Policies to fill in the gap reportedly include one plan to reduce emissions by 3 million tons by training German drivers to drive more economically.36 As noted by the EC Report cited above, the effectiveness of public education programs such as this may be problematic.

Harmonizing NAPs

The EU-ETS system involves an interplay between definitions and procedures that are EU-wide and those that are nationwide. The groundwork for the system is the Kyoto Protocol, which (1) defines the pollutants and sets the countries' emissions targets; (2) defines the scope of participation: Annex 1 countries may implement emissions trading programs, and non-Annex 1 countries may participate through the CDM; (3) defines baseline emissions years and sinks; and (4) sets national inventory and compliance requirements. Within this framework, the EU defines the elements that make the EU-ETS work, including industry participants, the unit of trade (tradeable allowances equal to 1 metric ton of CO2), trading periods, settling up procedures, and linkages within and beyond the EU.

With respect to the individual Member's NAPs, the EC harmonizes the NAPs with respect to penalties, allocation method (e.g., grandfathering), monitoring, and registries with the goal of achieving the Kyoto targets. It allows Members flexibility with respect to allocations to individual participants, the extent to which banking is permitted, and whether to permit the auctioning of up to 10% of allowances in the second trading period. As a result of this framework, there are significant differences between Members with respect to participant definitions, industry level emissions caps and allocations, and enforcement.

To increase the economic and administrative efficiency of the ETS, some stakeholders are interested in improved harmonization of NAPs by the EC. Besides the issue of new entrants and definition of affected units specifically identified by the EC survey, harmonization issues include allocation methods and the use of auctions, the degree to which JI and CDM credits may be used for compliance, and monitoring, verification, and reporting rules. Table 5 illustrates the scope of potential harmonization issues facing the EU-ETS.

Table 5. NAP Harmonization Issues

Subject	Source of Member Differences
Definition of allowances	Financial and tax treatment of allowances
Reliable emissions inventories	Inventory standards
Banking of allowances	Whether and how much banking permitted
Emissions caps	Stringency of caps and the extent of JI and CDM credits permitted
Monitoring, verification and reporting	Procedures and processes
Allocation	Allocation methods and whether and how much auctioning permitted
National registries	Design details
Voluntary participants	Whether to allow pooling or opt-in/opt-out
Definition of mandatory participants	Definition of sectors, size, installation, new entrant, and treatment of closures

Source: Adapted from Fiona Mullins, *EU ETS Implementation: Room for Harmonisation* (The Royal Institute of International Affairs, 2005).

One issue of particular interest is the effort to increase the use of "benchmarking" standards in setting allocations. Benchmarking generally involves allocating allowances based on best available technology and practices, rather than on historical emissions. However, the EC does not have the authority to scrutinize allocations at the facility level, so any allocation harmonization would be on a voluntary basis ("soft harmonization"). Also, allocation schemes, such as benchmarking, may not be suitable for some industries.[37] The EC's survey of ETS stakeholders revealed that although more than two-thirds of the respondents from the cement, aluminum, and chemical industries thought benchmarking was an "interesting alternative," less than a third of the respondents from the pulp and paper industry and refineries thought so.[38]

In many ways, diversity between Member countries with respect to the ETS is inevitable. As stated by Grubb and Neuhoff with respect to allocation:

> The final way in which the EU ETS differs from many other trading systems is in the devolution of allocation responsibilities, in this case to its 25 Member States. This was an essential part of the deal that enabled the adoption of the Directive: Member States would have never ceded to the European Commission the power to distribute valuable assets to their industries. Nor is the EU ETS unique in devolving powers of allocation: it is typical in a number of US systems. Moreover, there are different degrees of harmonization, applicable to different aspects of the EU ETS, and the Commission can and does seek to increase the degree of harmonization through guidance notes.[39] [footnote omitted]

New Entrants

The economic value of allowances is nowhere more evident than in discussions of new entrants. Indeed, as noted above, Germany is proposing to permit new coal-fired powerplants built between 2008-2012 to opt-out of the ETS for the first 14 years of operation in order to encourage construction. In its survey of ETS stakeholders, the EC found that 85% of all respondents favor a harmonized approach to new entrants and closures. Nearly 75% believed that those allowances should be provided free.40 Likewise, an EU questionnaire conducted by the European Environmental Agency's Topic Centre on Air and Climate Change indicated that most Member States would welcome harmonization of the treatment of new entrants and closures across the EU.41

Analogous to the U.S. acid rain program, EU states have set up reserves to provide allowances to new entrants. In general, these allowances are provided free, as that is widely seen as helping boost new investment. However, the allocation methods developed by the Member States differ. Most states have yet to dip into their reserves for new entrants; however, the importance of the reserve will increase as the ETS enters its second, and eventually third, trading period. The manner in which new entrants receive allowances may have a significant effect on the long-term direction of investment — whether it is directed toward low-carbon opportunities or used to support continuation of current economic development irrespective of its carbon intensity. Allocating allowances according to output and not differentiating according to the carbon intensity of the project would provide an incentive to develop low carbon alternatives. An example provided by Grubb and Neuhoff:

> New entrant reserves should be based on output or capacity, and avoid differentiating according to the CO_2-intensity of the new investment. In particular, giving more to coal than gas plants rewards investment in new coal facilities, which would conflict with objectives to tackle climate change, increase the cost of future emissions reductions, and in the long run could lead to higher electricity prices. The damaging effects would be amplified if carbon-intensive new entrants not only receive free allowances for the period 2008-2012 but also receive promises for subsequent periods.[42]

The proposed treatment of coal-fired powerplants by the German Government indicates how difficult it will be to direct future investment toward low-carbon projects. However, it could be argued that the long-term success of the ETS and the EU's commitment to Kyoto and any subsequent agreements rests on such a redirection with respect to new entrants and long-term investment.

Definition of Affected Units

Another area in which several Member States would like more harmonization across the EU is the definition of a combustion installation.43 Concerns revolve around ambiguity in the current definition of a combustion installation and the number of small installations covered under the ETS. The ETS applies to energy activities for all sectors with combustion installations above 20 MW of thermal rated input, oil refineries, coke ovens, and, subject to size criteria, iron and steel, cement, lime, glass, ceramics, and pulp and paper facilities. However, Finland and Sweden opted to include small district heating installations with a rated thermal input below 20 MW.44 In contrast, as noted previously, Germany is attempting to have some planned coal-fired powerplants, which will be large producers of CO2, able to opt-out of the ETS for 14 years.

In addition to the consistency issue, small installations (between 20MW and 50MW) account for 30% (about 3,000) of the total facilities covered under the ETS, but a very small percentage of total CO2 emissions.45 Surveying 22 Member States, 36% of the covered installations produced less than 10,000 metric tons of CO2 annually.46 The somewhat weak emissions data available at the time of the EU questionnaire suggest that while three-quarters of all emissions are produced by the largest 7.5% of installations, the small installations (under 10,000 metric tons) are responsible for less than 1%. Whether the ETS should continue to cover the roughly 3,500 facilities under 10,000 metric tons annually remains a hotly debated issue.

Expansion of Coverage

In choosing a gradual, incremental approach to emissions trading, the EU is relying on other programs to control greenhouse gas emissions in other sectors, such as transportation. The difficulties the EC may encounter in not choosing a comprehensive approach to begin with is suggested by its survey of stakeholders.

The survey suggests that the future direction of the ETS in terms of increasing coverage is toward incrementally adding more economic sectors, rather than addressing the more complex issue of a comprehensive system. Based on the survey, the focus is currently on the chemical, aviation, and aluminum industries.47 Given that the number-one recommendation for future implementation of the ETS is to provide participants with a longer time frame for implementation, it is unclear when the ETS will become as comprehensive as the European Commission would like.

CONCLUSION

At first glance, the ETS would appear an effective vehicle for the EU to meet its Kyoto Protocol obligations during the second trading period. The anticipated deficit between the second trading period for the original 15 Member States can be covered by trading with the newer Member States that anticipate a surplus. In addition, potential CERs and ERUs from the CDM and JI respectively may help maintain limits on allowance costs. Table 6 provides one series of estimates of available allowances for the Kyoto Protocol's five-year compliance period. Obviously, not all these allowances may be available to the EU alone; other countries, such as Japan and Canada, may decide to incorporate emissions trading into their implementation strategies and acquire allowances from these sources. Yet, the totals suggest that all else being equal, the supply of allowances would be adequate.

However, there are other considerations. The availability of surplus credits created via JI and CDM is restricted by the EC linking directive that requires that such credits be "supplemental" to a country's domestic efforts. Each country is to spell out what "supplemental" means in its NAP for the second trading period. Individual countries are likely to define that term differently — restricting allowance trades and purchases in some countries.

Another consideration is the overall commitment of the Kyoto Protocol. As noted earlier, the ETS only covers a percentage of the overall greenhouse gas emissions in the various Member States of the EU. Some sectors not covered by the ETS may grow faster than sectors covered by the ETS, creating difficulties for compliance. In particular, the transportation area is already a major source of concern.

A final consideration for the ETS is its suitability for directing long-term investment toward a low-carbon future — the ultimate goal of any climate change program. It is too early to tell whether the ETS market signal and individual countries' NAPs will move investment in the appropriate direction. The early signs are not particularly encouraging, with the 2005-2008 NAPs producing an over-allocation of allowances, and one major Member State, Germany, attempting to direct its second NAP toward carbon-intensive, coal-fired electric generating facilities rather than low-carbon alternatives. Reluctance by countries to redirect their NAPs and an inconsistent price from the ETS make the long-term effect of the ETS uncertain.

Table 6. International Supply of Emissions Credits and Allowances (cumulative total 2008-2012, million metric tons CO_2)

Source of Supply	Low Estimate	High Estimate
Clean Development Mechanism (CERs)	680	1,200
Joint Implementation (ERUs	120	980
Surplus Kyoto Allowances from Eastern Europe, Russia, and the Ukraine (AAUs)		

Source: Compiled by Grubb and Neuhoff, "Allocation and Competitiveness in the EU Emissions Trading Scheme: Policy Overview," p. 20.

APPENDIX: NORWAY'S TRADING SYSTEM

Norway, a non-EU country, also has an emissions trading system that began operating on January 1, 2005. Norway's system covers 51 facilities in the energy and process sectors such as oil refining and iron and steel processing, and has several features in common with the EU ETS.48 These sectors account for about 10-15% of the country's emissions. Other parts of Norway's industry, particularly its offshore oil and gas sector, are covered by the country's carbon tax of almost 40 euro per metric tonne of CO2 — much higher than the anticipated allowance price under the trading program. The carbon tax is levied on about 64% of the country's CO2 emissions — about half the country's total greenhouse gas emissions.49 The first phase of the trading program covers the period 2005-2007, with a second, expanded phase to begin in 2008.

Developing its initial trading scheme independently of the EU, Norway's "Quota Commission" (created in 1998) stressed that the 2008 program be as comprehensive as possible, suggesting that the system could include close to 90% of the country's greenhouse gas emissions.50 To achieve this coverage, the Commission envisioned a hybrid system of quotas, depending on cost-effectiveness and practicality considerations. Arguing in favor of regulating CO2 emissions at the producer (upstream) level from mobile sources and some stationary sources, the Quota Commission states:

> For these emissions, regulation at the producer level will not create weaker incentives for reducing emissions than regulation at the consumer level, because the volume of emissions from a particular commodity is not dependent on technology. For those process emissions which are recommended for inclusion in the system, regulation by quotas should be imposed at the end-user level in cases where the processes originate with major industrial companies. For process emissions stemming from a series of small sources, e.g. emissions of N_2O from commercial fertilizers, regulation by quotas should be imposed on retailers, or importers in order to avoid unacceptably high system costs.[51]

The Commission's recommendation that "regulation by quotas be imposed in part on the producer, in part on the sales or import chain, and in part on the end-user"[52] may be overtaken by development of the EU ETS, which is likely to influence the future direction of Norway's trading system. The International Energy Agency (IEA) states:

> The government had indicated earlier that it would consider expanding the [trading] system from 2008 to include as many sources of emissions as practical through an upstream system allocating allowances to fossil fuel producers and importers. With the adoption of the EU ETS design features, Norway seems to move away from this option.[53]

ENDNOTES

1 Six gases are included under the Kyoto Protocol: carbon dioxide, methane, nitrous oxide, hydrofluorocarbons, perfluorocarbons, and sulfur hexafluoride. The United States has not ratified the Kyoto Protocol and, therefore, is not covered by its provisions. For more information on the Kyoto Protocol, see CRS Report RL30692, *Global Climate Change: The Kyoto Protocol*, by Susan Fletcher.

2 Norway, a non-EU country, also has instituted a CO_2 trading system (described in Appendix A). Various other countries and a state-sponsored regional initiative located in the northeastern United States involving several states are developing mandatory cap-and-trade system programs, but are not operating at the current time. For a review of these emerging programs, along with other voluntary efforts, see International Energy Agency, *Act Locally, Trade Globally* (2005).

3 P.L. 101-549, Title IV (Nov. 15, 1990).

4 More information, including relevant directives, on the EU-ETS is available on the European Union's website at [http://europa.eu.int/scadplus/leg/en/lvb/l28012.htm].

5 Pew Center on Global Climate Change, *The European Union Emissions Trading Scheme (EU-ETS): Insights and Opportunities* (no date), available at [http://www.pewclimate.org/docUploads/EU%2DETS%20White%20Paper%2Epdf].

6 Directive 2003/87/EC of the European Parliament and of the Council of 13 October 2003 establishing a scheme for greenhouse gas emissions allowance trading within the Community and amending Council Directive 96/61/EC.

7 Commission of the European Communities, *Green Paper on Greenhouse Gas Emissions Trading within the European Union* (presented by the Commission), Brussels, COM(2000) 87 final (Mar. 8, 2000), p. 10.

8 The first political decision noted by the authors was the exclusion of process emissions from the chemical industry from the ETS. Catherine Boemare and Philippe Quirion, *Implementing Greenhouse Gas Trading in Europe: Lessons from Economic Theory and International Experiences*, Centre International de Recherche sur l'Environnement et le Developpement, CNRS/EHESS, France (June 2002), p. 5.

9 Commission of the European Communities, *Green Paper on Greenhouse Gas Emissions Trading with the European Union* (presented by the Commission), Brussels, COM(2000) 87 final (Mar. 8, 2000), p. 10.

10 Commission of the European Communities, *Green Paper on Greenhouse Gas Emissions Trading with the European Union* (presented by the Commission), Brussels, COM(2000) 87 final (Mar. 8, 2000), p. 13

11 International Energy Agency, *Act Locally, Trade Globally (*OCED/IEA, 2005), p. 74

12 Commission of the European Communities, Directive 2003/87/EC, available at [http://ec.europa.eu/environment/climat/emissions_plans.htm]

[13] Four of the 25 Member States (Cyprus, Luxembourg, Malta, and Poland) have not submitted information because their allowance registries are not operational yet. Cyprus and Malta do not have emissions targets under the Kyoto Protocol.

[14] European Commission, *EU Emissions Trading Scheme Delivers First Verified Emissions Data for Installations* (Brussels, 15 May 2006), available at[http://ec.europa.eu/comm/environment/climat/pdf/citl_pr.pdf]

[15] Comments of EU Environment Commissioner Stavros Dimas, as reported by Jeff Mason in "EU's Dimas Says States Allocated too Much CO_2 in '05" (Reuters, May 22, 2006).

[16] CAN Europe, *National Allocation Plans 2005-7: Do They Deliver? Summary for Policymakers* (April 2006), p. 2.

[17] European Commission, *EU Emissions Trading Scheme Delivers First Verified Emissions Data for Installations* (Brussels, May 15, 2006), available at [http://europa.eu.int/luxembourg/docs/217-2006_en.pdf]

[18] Point Carbon, *Carbon 2006: Towards a Truly Global Market* (28 February 2006) pp. 15-16. It should be noted that there is significant uncertainty in the estimates of the bilateral market.

[19] Margus Kaasik (CFO Eesti Energia), *Carbon Market: EU ETS* (May 9, 2006). Implied volatility is a measure used primarily in options analysis to estimate how much the market expects an asset price to move for an option price. Two hundred-fifty days is commonly used for this analysis, as it represents the number of business days in a year.

[20] Margus Kaasik (CFO Eesti Energia), *Carbon Market: EU ETS* (May 9, 2006), p. 13.

[21] Vivian E. Thomson in collaboration with the Pew Center on Global Climate Change, *Early Observations on the European Union's Greenhouse Gas Emissions Trading Scheme: Insights for United States Policymakers* (Apr. 19, 2006), p. 16.

[22] Point Carbon, *Carbon 2006: Towards a Truly Global Market* (Feb. 28, 2006) pp. 18-19.

[23] Annex 1 countries are the 36 industrialized countries and economies-in-transition listed in Annex 1 of the United Nations Framework Convention on Climate Change and allowed to engaged in JI and CDM projects under the Kyoto Protocol.

[24] The time between submission of a CDM project design document and approval can be as much as 18 months because CDM credits must be approved by the Designated National Authority (DNA) and the CDM Executive Board. Brown Rudnick Berlack Israels LLP, *Emissions Trading: Questions and Answers* (February 2006), p. 7.

[25] For a list of countries with carbon funds, see [http://carbonfinance.org/Router.cfm?Page=Funds&ItemID=24670]

[26] As reported in *Carbon Positive*, "CER Prices Stabilise After EU Market Hit" (June 14, 2006), available at [http://www.carbonpositive.net/viewarticle.aspx?articleID=137].

[27] European Commission, *Review of EU Emissions Trading Scheme* (November 2005).

[28] Two EU countries, Malta and Cyrus, are not included here because they are non-Annex 1 countries and, therefore, do not have mandatory reduction requirements under the Kyoto Protocol.

[29] EurActiv.com, *Transport to Stay Out of CO_2 Trading until 2013* (June 21, 2006).

[30] European Commission, *Implementing the Community Strategy to Reduce CO_2 Emissions from Cars: Fifth Annual Communication on the Effectiveness of the Strategy* (Brussels, June 22, 2005).

[31] Council conclusions of June 25, 1996.

[32] European Federation for Transport and Environment, *Cleaner is Cheaper* (Brussels, 2005). p. 1.

[33] European Commission, *Implementing the Community Strategy to Reduce CO_2 Emissions from Cars: Fifth Annual Communication on the Effectiveness of the Strategy* (Brussels, June 22, 2005) p. 7.

[34] On a like-kind basis. The NAP II cap is actually 482 million metric tons because it include 11 million tons for facilities covered under the second phase that were not covered under the first phase.

[35] Karoline Rogge, Joachim Schleich, Regina Betz, and Jos Cozijnsen, *Increasing the Ambition of EU Emissions Trading* (June 2006), p. 1.

[36] Roger Harrabin, BBC environmental analyst, *Germany to Spark "Climate Crisis"* (June 28, 2006).

[37] For more, see Michael Grubb and Karsten Neuhoff, "Allocation and Competitiveness in the EU Emissions Trading Scheme: Policy Overview," 6 *Climate Policy* (2006) pp. 7-30.

[38] European Commission, *Review of EU Emissions Trading Scheme: Survey Highlights* (November 2005), p. 15.

[39] Michael Grubb and Karsten Neuhoff, "Allocation and Competitiveness in the EU Emissions Trading Scheme: Policy Overview," 6 *Climate Policy* (2006), p. 17.

[40] European Commission, *Review of EU Emissions Trading Scheme: Survey Highlights* (November 2005), pp. 18-19.

[41] European Environment Agency's Topic Centre for Air and Climate Change, *Application of the Emissions Trading Directive by EU Member States*, (EEA Technical Report No. 2/2006, 2006), p. 30.

[42] Michael Grubb and Karsten Neuhoff, "Allocation and Competitiveness in the EU Emissions Trading Scheme: Policy Overview" 6 *Climate Policy* (2006), p. 22.

[43] European Environment Agency's Topic Centre for Air and Climate Change, *Application of the Emissions Trading Directive by EU Member States* (EEA Technical Report No. 2/2006, 2006), p. 30.

[44] These are generally owned by larger facilities which operate several installations covered by the ETS. See European Environment Agency's Topic Centre for Air and Climate Change, *Application of the Emissions Trading Directive by EU Member States* (EEA Technical Report No. 2/2006, 2006), p. 16.

[45] The installation number is for 22 Member States (Poland did not provide sufficient data). Emissions estimates were provided by only 14 States. Based on those 14 States, the allowance allocations to small installation accounted for about 2% of the total. See European Environment Agency's Topic Centre for Air and Climate Change, *Application of the Emissions Trading Directive by EU Member States* (EEA Technical Report No. 2/2006, 2006), p. 15.

[46] European Environment Agency's Topic Centre for Air and Climate Change, *Application of the Emissions Trading Directive by EU Member States* (EEA Technical Report No. 2/2006, 2006).

[47] European Commission, *Review of EU Emissions Trading Scheme* (November 2005), p. 11.

[48] International Energy Agency, *Act Locally, Trade Globally* (2005), p. 102.

[49] Ministry of the Environment, *Report 54 to the Storting (2000-2001): Norwegian Climate Policy*, at [http://odin.dep.no/md/engelsk/publ/stmeld/022001-040012/index-dok000-b-n-a.html]

[50] Ministry of the Environment, Trading in Greenhouse Gases (press release, Dec. 17, 1999), at [http://www.odin.no/odinarkiv/english/bondevik_I/md/022001-990070/dok-bn.html].

[51] The Quota Commission, *A Quota System for Greenhouse Gases* (Dec. 17, 1999), p.7, at [http://odin.dep.no/odinarkiv/norsk/dep/md/1999/eng/022021-220003/index-dok000-b-n-a.html]

[52] Ibid., p. 19.

[53] International Energy Agency, *Act Locally, Trade Globally* (2005), p. 103.

In: Global Climate Change Revisited
Editor: Horace M. Karling, pp. 83-102
ISBN 978-1-59454-039-4

Chapter 5

GLOBAL CLIMATE CHANGE*

John R. Justus and Susan R. Fletcher

SUMMARY

There is concern that human activities are affecting the heat/energy-exchange balance between Earth, the atmosphere, and space, and inducing global climate change, often termed "global warming." Human activities, particularly the burning of fossil fuels, have contributed to increased atmospheric carbon dioxide (CO_2) and other trace greenhouse gases. If these gases continue to accumulate in the atmosphere at current rates, most scientists believe significant global warming would occur through intensification of Earth's natural heat-trapping "greenhouse effect." Possible impacts might be seen as both positive and negative, depending on regional or national variations.

A warmer climate would probably have far-reaching effects on agriculture and forestry, managed and unmanaged ecosystems, including natural habitats, human health, water resources, and sea level, depending on climate responses. Although causal relationships between projected long-range global climate trends and record-setting warmth and severe weather events of the past two decades have not been firmly established, attention has been focused on possible extremes of climate change and the need for better understanding of climate processes to improve climate model projections.

The basic policy question remains: Given scientific uncertainties about the magnitude, timing, rate, and regional consequences of potential climatic change, what are the appropriate responses for U.S. and world decision makers?

Fossil-fuel combustion is the primary emissions, and also emits other source of CO_2 "greenhouse" gases. Because the U.S. economy is so dependent upon energy, and so much of U.S. energy is derived from fossil fuels, reducing these emissions poses major challenges and controversy.

The 1992 United Nations Framework Convention on Climate Change (UNFCCC), which the United States has ratified, called for a "non-binding" voluntary aim for industrialized countries to control atmospheric concentrations of greenhouse gases by stabilizing their

* Excerpted from CRS Report IB89005 dated May 12, 2006.

emissions at 1990 levels by the year 2000. The 1997 Kyoto Protocol to the UNFCCC goes further, and commits the major industrialized nations that have ratified it to specified, legally binding emissions reductions. On February 16, 2005, the Kyoto Protocol entered into force. According to the UNFCCC Secretariat, as of February 6, 2006, 160 nations and economic regional integration organizations had ratified the Protocol. The European Union instituted its emissions trading system under the Protocol at the beginning of 2005.

In March 2001, the Bush Administration rejected the Kyoto Protocol, and thus the United States is not party to it (and therefore is not subject to its requirements) as it enters into force. President Bush concluded a cabinet-level climate policy review with an announcement in 2002 of a "new approach" for the United States based on reducing the greenhouse gas intensity (greenhouse gas emissions per unit of GDP) of the U.S. economy.

This report briefly reviews the status of climate science, international negotiations, and congressional activity focused specifically on climate change.

Most Recent Developments

On February 16, 2005, the Kyoto Protocol to the United Nations Framework Convention on Climate Change (UNFCCC) entered into force, committing those nations that have ratified it to specified mandatory reductions in greenhouse gas emissions below 1990 levels. The 160 nations that had ratified the Protocol as of February 6, 2006, now represent 61.6% of the 1990 emissions baseline among developed nations (55% must be accounted for by industrialized countries that have ratified in order for the Protocol to enter into force). The United States is a party to the UNFCCC, but not to the Kyoto Protocol. The first meeting of the Kyoto Protocol parties occurred November 28-December 9, 2005, in a joint meeting with the 11th Conference of the Parties (COP-11) of the UNFCCC.

On June 22, 2005, the U.S. Senate passed a "Sense of the Senate" resolution as an amendment to H.R. 6, the Energy Policy Act of 2005, which included findings that greenhouse gases are accumulating in the atmosphere, causing average temperatures to rise, and there is "growing scientific consensus that human activity is a substantial cause of greenhouse gas accumulation in the atmosphere." It stated the sense of the Senate "that Congress should enact a comprehensive and effective national program of mandatory, market-based limits and incentives on emissions of greenhouse gases that slow, stop, and reverse the growth of such emissions... ," while not significantly harming the U.S. economy, and in a way that will encourage comparable action by other nations. This resolution was not included in the final energy bill by the conference committee.

Following suit, on May 10, 2006, the House Committee on Appropriations in the draft FY2007 Interior-Environment Appropriations bill adopted sense-of-the-Congress language stating that there should be enacted a comprehensive and effective national program of mandatory, market-based limits and incentives on emissions of greenhouse gases that slow, stop, and reverse the growth of such emissions. That language calls for limits and incentives that do not significantly harm the U.S. economy and that will encourage comparable action by our major trading partners and key contributors to global emissions. Also adopted was language directing EPA to study the ways global climate change will affect human health and to recommend ways the nation should improve medical preparedness and response capabilities and public health systems to deal with the health impacts associated with increased pollution and other environmental changes.

BACKGROUND AND ANALYSIS

Global Climate Change Science

A large number of scientists believe that human activities, which have increased atmospheric concentrations of carbon dioxide (CO_2) by 35% from preindustrial values of 280 parts per million (ppm) to 378 ppm over the past 150 years, are leading to an increase in global average temperatures. Global temperatures have already risen 0.6°C (0.9°F) in the last 100 years, and, according to model projections, might rise anywhere from as little as 1.8°C to as much as 7.1°C (2.7°F to 10.7°F) over the next 100 years. However, the science of "global warming" is not without challengers, who argue that scientific proof is incomplete or contradictory, and that there remain many uncertainties about the nature and direction of Earth's climate. Nevertheless, there is significant concern that human activities, such as the burning of fossil fuels, industrial production, deforestation, and certain land-use practices are increasing atmospheric concentrations of carbon dioxide (CO_2) that, along with increasing concentrations of other trace gases such as methane (CH_4), nitrous oxide (N_2O), hydrofluoro-carbons (HFCs), perfluorocarbons (PFCs), and sulfur hexafluoride (SF_6), may be leading to changes in the chemical composition and physical dynamics of Earth's atmosphere, including how heat/energy is distributed between the land, ocean, atmosphere and space.

Greenhouse Gases: Sources and Trends

Scientists have found that the four most important variable greenhouse gases, whose atmospheric concentrations can be influenced by human activities, are carbon dioxide (CO_2), methane (CH_4), nitrous oxide (N_2O), and chlorofluorocarbons (CFCs). Historically, CO_2 has been the most important, but those other atmospheric trace gases are also radiatively active, in that they can affect Earth's heat budget and thereby contribute to a greenhouse warming of the lower atmosphere.

The amount of carbon cycling from naturally occurring processes each year through the biosphere as CO_2 is enormous — some 800 billion tons. Ice cores and other proxy climate data, which also indicate CO_2 concentrations in the atmosphere, have shown, in general, a relatively stable global climate, at least over the past 10,000 years. As such, many scientists suggest that the amount of CO_2 generated by natural processes is about equal to the amounts absorbed and sequestered by natural processes. However, human activity since the Industrial Revolution (c.a. 1850), and primarily in the form of burning fossil fuels, is now generating some additional 24 billion tons of CO_2 per year. Available evidence shows that about half this amount is absorbed by natural processes on land and in the ocean, and that atmospheric concentrations of CO_2 are currently about 35% higher at 378 parts per million (ppm) than they were some 150 years ago at 280 ppm. Some scientists believe that a large amount of CO_2 may be stored in northern latitude soils and in temperate and tropical forests, suggesting a greater importance of the role of natural resources management and land-use practices in these regions, including burning of biomass and deforestation. Scientists estimate that anthropogenic emissions of CO_2 alone may account for as much as a 60% increase in global mean temperatures of 0.9oF, since 1850.

The combined radiative forcing from the other trace gases is approximately equal to that of CO_2 and, collectively, they are projected to contribute about as much to potential global warming over the next 60 years as CO_2. Some of the halogenated compounds, while present in the atmosphere at very low concentrations, are of considerable interest because of their high global warming potentials (GWPs) and long atmospheric residence times. Methane concentrations had been rising from a preindustrial value of around 700 parts per billion (ppb) up to the mid-1990s, when they began to level off to a present value of about 1,766 ppb, a 152% increase. Nitrous oxide concentrations have been rising from a preindustrial value of around 270 parts per billion (ppb) up to a present value of about 317 ppb, representing a 17% increase. The atmospheric concentrations of the chlorofluorocarbons CFC-11, at 268 parts per trillion (ppt), and CFC 113, at 84 ppt, have been on the decline in recent years in response to the Montreal Protocol of the late 1980's. The concentration of CFC-12 at 533 ppt is also anticipated to decline, but the decline is expected to be delayed due to (1) a longer atmospheric lifetime than CFC-11 or CFC-113, (2) its use in long-lasting appliances such as home refrigerators, and, possibly, (3) particularly extensive stockpiling at the global scale, due to a perceived lack of suitable replacement at the time production decreases were being mandated. The Kyoto Protocol on Climate Change also regulates three other trace gases: hydrofluorocarbons (HFCs), perfluorocarbons (PFCs), and sulfur hexafluoride (SF_6), whose limited concentrations in the atmosphere are anticipated to grow over the long term. Sulfate aerosols, a byproduct of air pollution, and other natural phenomena are also viewed as important for their transient and regional "climate cooling" effects in Earth's atmosphere.

On a related note, NASA scientist James Hansen has suggested that climate change benefits could be achieved through near-term regulation of non-CO_2 greenhouse gases. He proposed that reducing emissions of halocarbons (refrigerants), methane, nitrogen oxides, and carbon-black aerosols (soot) could have the effect of reducing ozone, itself a greenhouse gas, in the troposphere. Non-CO_2 greenhouse gases have relatively short atmospheric lifetimes compared with CO_2; however, most have a much larger global warming potential (gwp). This would suggest that controlling emissions of these greenhouse gases could reduce the rate and overall amount of potential climate warming from greenhouse gases, leaving only that projected from long-term CO_2 emissions whose full effects might not be realized for another 75-100 years. Nevertheless, Hansen emphasized that any actions to reduce emissions of these gases would need to be taken concomitantly with long-term strategies to reduce CO_2. Hansen also noted that modest gains from reducing CO_2 and non-CO_2 emissions in the near term could be achieved primarily through cleaner energy production.

Climate System Response

The most recent runs of state-of-the-art computer models of the Earth's climate (general circulation models — GCMs) have projected a globally averaged warming ranging from almost 3 to10.7degrees F over the next 100 years, if greenhouse gases continue to accumulate in the atmosphere at the current rate. Many climate scientists believe that such a warming could shift temperature zones, rainfall patterns, and agricultural belts and, under certain scenarios, cause sea level to rise. With regard to sea level rise, over the last 100 years, the global sea level has risen by about 10 to 25 centimeters (4 to 10 inches). It is likely that much of the rise in sea level has been related to the concurrent rise in global temperature over the last 100 years. On this time scale, the warming and the consequent thermal expansion of the world ocean may account for about 0.8 to 3 inches of the observed sea level rise, while the observed retreat of glaciers and ice caps may account for about 0.8 to 2 inches. Other factors are more difficult to quantify, including isostatic effects associated with rising or subsiding coastlines, neotectonics, and sedimentation. The rate of observed sea level rise suggests a net positive contribution from the huge ice sheets of Greenland and Antarctica, but observations of the ice sheets do not yet allow meaningful quantitative estimates of their separate contributions. The ice sheets remain a major source of uncertainty in accounting for past changes in sea level because of insufficient data about them over the last 100 years. Taking into account the ranges in the estimate of climate sensitivity and ice melt parameters, and the full set of greenhouse gas emissions scenarios, the models project an increase in global mean sea level of between 5 and 37 inches by the year 2100. In addition, because of the thermal inertia of the oceans, sea level could continue to rise for many centuries beyond 2100 even if concentrations of greenhouse gases were stabilized at that time. Some climate scientists further project that global warming could have far-reaching effects — some positive, some negative, depending how it may be experienced in a given region — on natural resources; ecosystems; food and fiber production; energy supply, use, and distribution; transportation; land use; water supply and control; and human health.

Skeptics of the global warming theory have called into question the reliability of the computer climate models and their output used to make projections of future warming that supported Kyoto Protocol negotiations. They also challenge some scientists' assertions that recent episodic weather events may seem more extreme in nature, and that this may be indicative of long-term climate change.

Evidence of natural variability of climate is large enough that even the record-setting warmth at the end of the 20th century has made it difficult for many climate scientists to state beyond a reasonable doubt that weather extremes experienced over the past two decades are attributable to "global warming," at least at the present time. However, the warming trend at the surface appears to be continuing. In some cases, causal relationships between seasonal and inter annual climate variability and present-day severe weather events are beginning to be recognized and even predicted, owing to an improved ability to observe such phenomena as *El Nino* and *La Nina,* the North Atlantic Oscillation (NAO), and the Pacific Decadal Oscillation (PDO). That notwithstanding, singular extreme weather events have focused public, academic, and government attention on possible outcomes of potential long-term climate change and a need for a better understanding of regional climates on decadal to century time scales.

National Oceanic and Atmospheric Administration's (NOAA) researchers reported that the 12 warmest years (globally averaged) since historical records have been kept occurred in the past two decades, with 1990 and 1998 among the warmest. Those records reveal that the year 2004 was the fourth warmest worldwide since records began. At least some of this warming, they concluded, is human-induced. On the other hand, satellite instruments — which, through indirect methods, measure the average temperature of the atmosphere in a deep column above the surface — are hard pressed to demonstrate any positive trends over the past 20-year period.

A report issued by the U.S. National Research Council's Board on Atmospheric Sciences and Climate, *Reconciling Observations of Global Temperature Change* (2000), attempted to resolve apparent disparities between temperature data measured at the surface and those from satellites. Scientists who question the surface temperature record state that such disparate trends invalidate the output of general circulation models (GCMs), many of which demonstrate homogenous warming throughout all the levels of the Earth's atmosphere. Panel scientists concluded that there may be a systematic disconnect between the upper and near surface atmosphere and cited physical processes not currently accounted for in GCMs that may have a unique impact on the upper atmosphere. In addition, they acknowledged that only long-term, systematic monitoring of the upper atmosphere could resolve the differences in temperature trends. Scientific work continues in the opposing communities and among government and university modeling centers, accompanied by analysis of weather balloon and radiosonde data and re-analysis of surface temperature and satellite data sets, with the prospect of advancing resolution of this pivotal issue. The most recent entry into this fray, released May 2, 2006, by the U.S. Climate Change Science Program (CCSP), is the first of what will be 21 Synthesis and Assessment Products, this one entitled *Temperature Trends in the Lower Atmosphere —Steps for Understanding and Reconciling Differences; CCSP Synthesis and Assessment Product 1.1* (see [http://www.climatescience.gov/Library/sap/sap1-1/finalreport/default.htm]).

The Intergovernmental Panel on Climate Change (IPCC), jointly established in 1988 by the United Nations World Meteorological Organization (WMO) and the United Nations Environment Program (UNEP), reported in its *Second Assessment* (1996) that "... [such] a change is unlikely to be entirely natural in origin ... [and that] the balance of evidence, from changes in global mean surface air temperature and from changes in geographical, seasonal, and vertical patterns of atmospheric temperature, suggests a discernible human influence on global climate." Issuing an updated conclusion in January 2001, the Intergovernmental Panel in its *Third Assessment* (2001), reported that a firmer association between human activities and climate seemed to have emerged, stating that "... in the light of new evidence and improved understanding, and taking into account the remaining uncertainties, most of the observed warming over the last 50 years is likely to have been due to the increase in greenhouse gas concentrations." That was news, because reservations about the source of the past century's warming and whether it bore a human fingerprint are often cited in policy debates, usually in support of deferring actions aimed at mitigating possible global warming. In addition, the IPCC reported a higher range of potential warming — roughly between 2.7 and just under 11 degrees Fahrenheit over the next 100 years.

In the United States, the national assessment report *Climate Change Impacts on the United States: The Potential Consequences of Climate Variability and Change* (2000), published and released under the auspices of the U.S. Global Research Program, received criticism from many of those involved in its review. Critics claimed that many of the model-projected impacts of possible future climate changes were overstated and unsubstantiated. The National Assessment Synthesis Team (NAST), with overall authority for the report, countered that much of the criticism it had received did not take into account the time scales upon which the report was based; the report targeted the effects of climate toward the middle of this century to the end of the next. Also, seemingly contradictory outcomes were produced by the two climate models selected for making the climate projections, casting some lingering doubt on the overall value and utility of the results for decision makers at the local, regional, and national levels. Various regional and resource-focused assessments are now available at the USGCRP website, [http://www.usgcrp.gov/usgcrp/nacc/default.htm]. A final synthesis report by the NAST, of the same title and consisting of an overview of all of the regional and sectoral studies, was released in December 2000.

On June 6, 2001, a Committee on the Science of Climate Change of the U.S. National Research Council (NRC) released a report, *Climate Change Science: An Analysis of Some Key Questions*, stating that global warming could well have serious societal and ecological impacts by the end of this century. Commissioned by the Bush White House and prepared by 11 of the nation's leading climate scientists, the report summarized the current state of knowledge on climate change and confirmed that the climatic changes observed during the past several decades were most likely due to human activities. The committee members warned, however, that they could not rule out the possibility that the climate's natural variability could be responsible for a significant portion of that trend. The authors agreed that human-induced warming and sea level rise were expected to continue through the 21st century and beyond, but they emphasized that current predictions of the magnitude and rate of future warming "should be regarded as tentative and subject to future adjustments (either upward or downward)."

The NRC report generally concurred with the latest conclusions of the Intergovernmental Panel on Climate Change, which found that the Earth warmed by about 1 degree Fahrenheit during the 20th century, and that most of the warming of the past 50 years was probably due to increases in greenhouse gas concentrations in the atmosphere. The full report, *Climate Change Science: An Analysis of Some Key Questions*, is available online at [http://fermat.nap.edu/html/climatechange/] or may be downloaded as a PDF file at [http://www.nap.edu/catalog.php?record_id=10139].

President Bush made a speech on global climate change from the Rose Garden on June 11, 2001, following release of that NRC *Key Questions* report and completion of a cabinet-level review of climate change options. In that speech, timed just before his trip to Europe to meet with leaders there, the President acknowledged that the world has warmed and that greenhouse gases have increased, largely due to human activity, but emphasized that the magnitude and rate of future warming are unknown.

The Policy Context

Since the conclusion of the United Nations Framework Convention on Climate Change in 1992, U.S. climate policy has been evolving through several different stages, first under the Clinton Administration, and then under a very different approach under the Bush Administration. This history is reviewed briefly below. (For more information on U.S. Climate Policy and how it has developed, see CRS Report RL31931, *Climate Change: Federal Laws and Policies Related to Greenhouse Gas Reductions.*)

Clinton Administration Policies

Taking office the year after the UNFCCC was completed, the Clinton Administration presided over early U.S. efforts to deal internationally with climate change, and to participate in formulation of the Kyoto Protocol to the UNFCCC. On October 19, 1993, President Clinton released his *Climate Change Action Plan* (CCAP), which proposed voluntary domestic measures to attain greenhouse gas emissions stabilization as outlined by the UNFCCC, to stabilize U.S. emissions at 1990 levels by the year 2000. The CCAP called for comprehensive voluntary measures by industry, utilities and other large-scale energy users. CCAP stressed energy-efficiency upgrades through new building codes in residential and commercial sectors, and other improvements in energy generating or using technologies. Large-scale tree planting and forest reserves were encouraged to enhance sequestration of carbon dioxide and to conserve energy. Other aspects of the plan addressed mitigation of greenhouse gases other than CO_2. The CCAP avoided mandatory command and control measures.

On November 12, 1998, President Clinton instructed a representative to sign the Kyoto Protocol to "lock in" U.S. interests achieved during negotiations. This act drew protest by some in Congress. Some Members claimed Clinton action was in violation of the June 1997 Byrd/Hagel Resolution (S.Res.98) that required an economic analysis of legally binding emission reductions on the United States, as well as binding obligations for all UNFCCC parties, including developing countries. The President announced he would continue to pursue"meaningful" commitments from key developing countries before he would send the treaty to the Senate for advice and consent.

The Clinton Administration released an economic analysis in July 1998, prepared by the Council of Economic Advisors, that concluded that with emissions trading among the Annex B/Annex I countries, and participation of key developing countries in the "Clean Development Mechanism" — which grants the latter business-as-usual emissions rates through 2012 — the costs of implementing the Kyoto Protocol could be reduced as much as 60% from many estimates. Other economic analyses, however, prepared by the Congressional Budget Office and the DOE Energy Information Administration (EIA), and others, projected a potentially large declines in GDP from implementing the Protocol.

On November 11, 2000, President Clinton issued a statement on "Meeting the Challenge of Global Warming" in response to the results of the report *Climate Change Impacts on the United States: The Potential Consequences of Climate Variability and Change* (available at [http://www.usgcrp.gov/usgcrp/nacc/default.htm]). In his statement, President Clinton said he would promulgate new regulations for U.S. electric power plants, imposing emissions caps on sulphur, nitrogen oxides, mercury, and CO_2. He also called for establishment of a domestic emissions trading program and promised a continued U.S. leadership role in climate change to set an example for other industrialized countries. President Clinton announced he would take such steps as necessary to keep the United States on target for meeting Kyoto Protocol goals, if certain concessions were made regarding international adoption of flexible mechanisms such as emissions trading, the clean development mechanism (CDM), credit for carbon sinks, and accountable, legally-binding, compliance mechanisms.

Bush Administration Policies

Soon after taking office, the Bush Administration had asked for a delay in resumption of the collapsed COP-6 negotiations (see COP-6 discussion below), in order to allow time for consideration of its approach and policies. Talks were accordingly scheduled for the second half of July. However, in late March 2001, the Bush Administration rejected the Kyoto Protocol — causing widespread concern among the EU nations — citing lack of developing country participation and possible harm to the U.S. economy. This followed extensive press attention to, first, statements by the EPA Administrator that — pursuant to a campaign statement by then-candidate George W. Bush — carbon dioxide would be included in a multi-pollutant regulatory effort; and then a repudiation of that position and clarification by President Bush and Administration spokespersons that carbon dioxide would not be regulated.

President Bush made a policy statement in mid-June 2001, resulting from a continuing cabinet-level review of climate change options, in which he outlined the U.S. approach as rejecting the Kyoto Protocol and favoring voluntary actions, increased scientific research, and market mechanisms. This preceded his trip to Europe for meetings with European heads of state, which ended with statements that Europe and the United States "agree to disagree" on climate change approaches. President Bush also outlined a Climate Change Research Initiative (CCRI) and a National Climate Change Technology Initiative (NCCTI), along with a new Cabinet-level Committee on Climate Change Science and Technology Integration to oversee their implementation. The CCRI and the NCCTI focus on short-term, policy-relevant objectives of climate change science and climate change technology applications, respectively. A previously established U.S. Global Change Research Program (USGCRP) supports long-term, fundamental, scientific research objectives.

Both the new CCRI and the existing USGCRP were combined for the first time into the Climate Change Science Program (CCSP) in the FY2004 budget. The FY2007 budget requests a total spending level of $1.717 billion for research managed by the CCSP, which is $4 million (+0.2%) above the FY2006 funding estimate of $1.713 billion. Included in the $1.717 billion CCSP funds are $200 million for the CCRI. While funding for the embedded CCRI experienced growth over two fiscal years from FY2003 to FY2005, the FY2007 funding request for CCRI at $200 million is level with the FY2006 funding estimate of $200 million. That leaves the FY2007 funding request for the embedded USGCRP standing at $1.517 billion, which is $4 million (+0.3%) above the FY2006 funding estimate of $1.513 billion. Some $2.98 billion is in the requested FY2007funding profile for technology research and development in the Climate Change Technology Program(CCTP) and its embedded NCCTI, an amount $207 million (+7.5%) above the FY2006 funding level of $2.773 billion. The budget request emphasized that the Climate Change Technology Program Office was in the process of addressing the nearly 300 comments received on its *Draft Strategic Plan,* in advance of publishing a final *Strategic Plan on Climate Change Technology* later on in 2006.

Two issues of concern for Congress are the extent to which spending for the CCRI and CCTP represents new money versus how much is attributable to the reclassification of ongoing research and technology programs, and whether reduced funding in some cases, or virtual level funding in most other cases, might be deemed necessary or sufficient to accomplish the work of the CCSP and the CCTP. The Administration released a *Climate Change Science Program Strategic Plan* on July 24, 2003. The plan included five major research goals and dozens of specific research targets as well as 23 written synthesis and assessment products with deadlines. The plan did not give budget details. The National Research Council of the National Academy of Sciences conducted an independent review of the *Strategic Plan* and in April 2004 published its overall assessment in a 51-page report, *Implementing Climate and Global Change Research: A Review of the Final U.S. Climate Change Science Program Strategic Plan.*[1] To complement the *CCSP Strategic Plan*, the Department of Energy has released several guidance documents for Climate Change Technology Program activities:

Climate Change Technology Program Strategic Plan Public Review Draft on January 24, 2006, *U.S. Climate Change Technology Program: Technology Options for the Near and Long Term* on September 15, 2005, and *U.S. Climate Change Technology Program: Vision and Framework for Strategy and Planning*, on August 5, 2005. Release of the completed final CCTP strategic plan is expected sometime in 2006.

In June 2001, the Europeans announced their intentions to proceed with ratification of the Kyoto Protocol, while President Bush indicated the United States would continue to participate in negotiations of the UNFCCC parties in order to pursue its own objectives, but would not participate directly in Kyoto Protocol negotiations. When talks resumed among UNFCCC parties at "COP-6 resumed" in mid-July in Bonn, Germany, and continued in the fall of 2001 in Marrakech, Morocco at COP-7, the United States delegation did not make new proposals and declined to participate in negotiations on issues of the Kyoto Protocol. Agreement among the other parties was found on the remaining Protocol issues at COP-7, and they announced that they would seek ratification of the Kyoto Protocol, and its entry into force — even without the participation of the United States.

On February 14, 2002, apparently concluding the cabinet-level review of climate change underway since early 2001, President Bush announced a U.S. policy for climate change: a "new approach for meeting the long-term challenge of climate change." The centerpiece of this announcement was the plan to reduce greenhouse gas intensity of the U.S. economy by 18% over the next 10 years. Greenhouse gas intensity measures the ratio of greenhouse gas emissions to economic output, and has been declining in the United States over the past several years. The Administration stated that the goal, to be met through voluntary action, is to achieve efficiency improvements that would reduce the 183 metric tons of emissions per million dollars of gross domestic product (GDP) to 151 in 2012. The plan noted that "if, in 2012, we find that we are not on track toward meeting our goal, and sound science justifies further policy action, the United States will respond with additional measures that may include a broad, market-based program" and other incentives and voluntary measures to accelerate technology development.

In addition, the policy directed the Secretary of Energy in consultation with other key agencies, to "substantially improve the emission reduction registry" to upgrade the voluntary emission reduction program under section 1605(b) of the 1992 Energy Policy Act, to bring about enhanced measurement accuracy, reliability, and verifiability. Other measures included providing for protected, transferable emission reduction credits, increased funding of $700 million in total climate-related spending, and a new management structure to coordinate climate change and technology research. Domestic policies such as tax incentives for renewable energy and new technology, development of fuel-efficient vehicles and cleaner fuels, and carbon sequestration were also proposed, along with several international bilateral initiatives and relatively modest increases in foreign assistance.

Some observers praised the plan for taking a practical, conservative approach to government action and for relying on voluntary measures. Critics observed that voluntary approaches by themselves have not historically often been effective, and noted that the reductions in energy intensity are very little different from current trends and would allow for significant increases in overall greenhouse gas emissions rather than reductions.

Continuing to encourage voluntary action rather than mandatory requirements, the Administration detailed on February 12, 2003, a set of voluntary agreements by various industry groups under an umbrella initiative titled Climate VISION (Voluntary, Innovative Sector Initiatives: Opportunities Now). These initiatives by sectoral groups involve actions to reduce greenhouse gas emissions and improve energy efficiency. (For a fuller description, see [http://www.whitehouse.gov/news/releases/2002/02/climatechange.html].)

International Action

The United States was involved in negotiations and international scientific research on climate change prior to ratifying the 1992 U.N. Framework Convention on Climate Change (UNFCCC). This included passage of a National Climate Program Act of 1978 (P.L. 95-367). These activities are discussed in CRS Report RL30522, *Global Climate Change: A Survey of Scientific Research and Policy Reports*, which covers early aspects of the scientific debate and contains a chronology of U.S. government involvement in climate change policy before 1992.

U.N. Framework Convention on Climate Change (UNFCCC)

The United Nations Framework Convention on Climate Change (UNFCCC) was opened for signature at the 1992 United Nations Conference on Environment and Development (UNCED) conference in Rio de Janeiro (known by its popular title, the Earth Summit). On June 12, 1992, the United States, along with 153 other nations, signed the UNFCCC, that upon ratification committed signatories' governments to a voluntary "non-binding aim" to reduce atmospheric concentrations of greenhouse gases with the goal of "preventing dangerous anthropogenic interference with Earth's climate system." These actions were aimed primarily at industrialized countries, with the intention of stabilizing their emissions of greenhouse gases at 1990 levels by the year 2000; and other responsibilities would be incumbent upon all UNFCCC parties. The parties agreed in general that they would recognize "common but differentiated responsibilities," with greater responsibility for reducing greenhouse gas emissions in the near term on the part of developed/industrialized countries, which were listed and identified in Annex I of the UNFCCC and thereafter referred to as "Annex I" countries.

On September 8, 1992, then-President Bush transmitted the UNFCCC for advice and consent of the U.S. Senate to ratification. The Foreign Relations Committee approved the treaty and reported it (Senate Exec. Rept. 102-55) October 1, 1992. The Senate consented to ratification on October 7, 1992, with a two-thirds majority vote. President Bush signed the instrument of ratification October 13, 1992, and deposited it with the U.N. Secretary General. According to terms of the UNFCCC, having received over 50 countries' instruments of ratification, it entered into force March 24, 1994.

Since the UNFCCC entered into force, the parties have been meeting annually in conferences of the parties (COP) to assess progress in dealing with climate change, and beginning in the mid-1990's, to negotiate the Kyoto Protocol to establish legally binding obligations for developed countries to reduce their greenhouse gas emissions. After completion of the Protocol in 1997, COP meetings focused on formulating the operational rules that would prevail as nations attempted to meet their obligations to reduce emissions. These rules were essentially agreed upon at COP-7 (see below) in 2001. As of February 16, 2005, the Kyoto Protocol entered into force. At that time, 141 nations had ratified it, including 35 of the 38 Annex B industrialized countries. Those Annex B parties to the UNFCCC that have ratified the Kyoto Protocol continue to express hope that the United States will re-engage in international efforts to reduce greenhouse gas emissions. As of February 6, 2006, nations that have ratified or accepted the Kyoto Protocol number 160.

COP-1, the Berlin Mandate

The UNFCCC Conference of Parties met for the first time in Berlin, Germany in the spring of 1995, and voiced concerns about the adequacy of countries' abilities to meet commitments under the Convention. These were expressed in a U.N. ministerial declaration known as the "Berlin Mandate," which established a two-year Analytical and Assessment Phase (AAP), to negotiate a "comprehensive menu of actions" for countries to pick from and choose future options to address climate change which for them, individually, made the best economic and environmental sense. The Berlin Mandate exempted non-Annex I countries from additional binding obligations, in keeping with the principle of "common but differentiated responsibilities" established in the UNFCCC — even though, collectively, the larger, newly industrializing countries were expected to be the world's largest emitters of greenhouse gas emissions 15 years hence.

COP-2, Geneva, Switzerland

The Second Conference of Parties to the UNFCCC (COP-2) met in July 1996 in Geneva, Switzerland. Its Ministerial Declaration was adopted July 18, 1996, and reflected a U.S. position statement presented by Timothy Wirth, former Under Secretary for Global Affairs for the U.S. State Department at that meeting, which (1) accepted the scientific findings on climate change proffered by the Intergovernmental Panel on Climate Change (IPCC) in its second assessment (1995); (2) rejected uniform "harmonized policies" in favor of flexibility; and (3) called for "legally binding mid-term targets."

COP-3, the Kyoto Protocol on Climate Change

The Kyoto Protocol to the United Nations Framework Convention on Climate Change was adopted by COP-3, in December 1997 in Kyoto, Japan, after intensive — and tense — negotiations. Most industrialized nations and some central European economies in transition (all defined as Annex B countries) agreed to legally binding reductions in greenhouse gas emissions of an average of 6%-8% below 1990 levels in the years 2008-2012, defined as the first emissions budget period. The United States would be required to reduce its total emissions an average of 7% below 1990 levels. (For more details, see CRS Report RL30692, *Global Climate Change: The Kyoto Protocol.*) The Protocol provides that it will enter into force when it has been ratified by 55 countries, accounting for 55% of developed country emissions in 1990. In 1997, prior to the completion of the Protocol, the U.S. Senate passed S.Res. 98, which urged the President not to agree to a treaty that did not include binding commitments for developing countries, or that would cause harm to the U.S. economy. As noted above, although President Clinton did sign the Protocol in 1998, it was never submitted by his administration to the Senate because it would not have met the conditions of S.Res. 98.

The Clinton Administration initiated funding efforts to address climate change; in the FY2001 budget request funding was included for a Climate Change Technology Initiative (CCTI) first introduced in his FY1999 budget. Somewhat reduced funding for the climate technology initiatives was received in previous years.

COP-4, Buenos Aires

COP-4 took place in Buenos Aires in November 1998. It had been expected that the remaining issues unresolved in Kyoto would be finalized at this meeting. However, the complexity and difficulty of finding agreement on these issues proved insurmountable, and instead the parties adopted a two-year "Plan of Action" to advance efforts and to devise mechanisms for implementing the Kyoto Protocol, to be completed by 2000.

COP-5, Bonn, Germany

The 5th Conference of Parties to the U.N. Framework Convention on Climate Change met in Bonn, Germany, between October 25 and November 4, 1998. It was primarily a technical meeting, and did not reach major conclusions.

COP-6, the Hague, Netherlands

When COP-6 convened November 13-25, 2000, in The Hague, Netherlands, discussions evolved rapidly into a high-level negotiation over the major political issues. These included major controversy over the United States' proposal to allow credit for carbon "sinks" in forests and agricultural lands, satisfying a major proportion of the U.S. emissions reductions in this way; disagreements over consequences for non-compliance by countries that did not meet their emission reduction targets; and difficulties in resolving how developing countries could obtain financial assistance to deal with adverse effects of climate change and meet their obligations to plan for measuring and possibly reducing greenhouse gas emissions. In the final hours of COP-6, despite some compromises agreed between the United States and some EU countries, notably the United Kingdom, the EU countries as a whole, led by Denmark and Germany, rejected the compromise positions, and the talks in The Hague collapsed. Jan Pronk, the President of COP-6, suspended COP-6 without agreement, with the expectation that negotiations would later resume. It was later announced that the COP-6 meetings (termed "COP-6 bis") would be resumed in Bonn, Germany, in the second half of July. The next regularly scheduled meeting of the parties to the UNFCCC — COP-7 — had been set for Marrakech, Morocco, in October-November 2001. (For more detailed discussion of COP-6 issues, see CRS Report RL30692, *Global Climate Change: The Kyoto Protocol.*)

COP-6 "bis," Bonn, Germany

When the COP-6 negotiations resumed July 16-27, 2001, in Bonn, Germany, little progress had been made on resolving the differences that had produced an impasse in The Hague. However, this meeting took place after President George Bush had become the U.S. President, and had rejected the Kyoto Protocol in March; as a result the United States delegation to this meeting declined to participate in the negotiations related to the Protocol, and chose to act as observers. As the other parties negotiated the key issues, agreement was reached on most of the major political issues, to the surprise of most observers given the low level of expectations that preceded the meeting. The agreements included:

(1) Mechanisms — the "flexibility" mechanisms which the United States had strongly favored as the Protocol was initially put together, including emissions trading; joint implementation; and the Clean Development Mechanism (CDM), which provides funding from developed countries for emissions reduction activities in developing countries, with credit for the donor countries. One of the key elements of this agreement was that there would be no quantitative limit on the credit a country could

claim from use of these mechanisms, but that domestic action must constitute a significant element of the efforts of each Annex B country to meet their targets.

(2) Carbon sinks — credit was agreed to for broad activities that absorb carbon from the atmosphere or store it, including forest and cropland management, and revegetation, with no overall cap on the amount of credit a country could claim for sinks activities. In the case of forest management, an Appendix Z establishes country-specific caps for each Annex I country, for example, a cap of 13 million tons could be credited to Japan (which represents about 4% of its base-year emissions). For cropland management, countries could receive credit only for carbon sequestration increases above 1990 levels. (3) Compliance — final action on compliance procedures and mechanisms that would address noncompliance with Protocol provisions was deferred to COP-7, but broad outlines of consequences for failing to meet emissions targets would include a requirement to "make up" shortfalls at 1.3 tons to 1; suspension of the right to sell credits for surplus emissions reductions; and a required compliance action plan for those not meeting their targets. (4) Financing — three new funds were agreed upon to provide assistance for needs associated with climate change; a least-developed-country fund to support National Adaptation Programs of Action; and a Kyoto Protocol adaptation fund supported by a CDM levy and voluntary contributions.

A number of operational details attendant upon these decisions remained to be negotiated and agreed upon; these were the major issues of the COP-7 meeting that followed.

COP-7, Marrakech, Morocco

At the COP-7 meeting in Marrakech, Morocco, October 29-November 10, 2001, negotiators in effect completed the work of the Buenos Aires Plan of Action, finalizing most of the operational details and setting the stage for nations to ratify the Protocol. The United States delegation continued to act as observers, declining to participate in active negotiations. Other parties continued to express their hope that the United States would re-engage in the process at some point, but indicated their intention to seek ratification of the requisite number of countries to bring the Protocol into force (55 countries representing 55% of developed country emissions of carbon dioxide in 1990). A target date for bringing the Protocol into force was put forward — the August-September 2002 World Summit on Sustainable Development (WSSD) to be held in Johannesburg, South Africa — but this target was not met. The main decisions at COP-7 included operational rules for international emissions trading among parties to the Protocol and for the CDM and joint implementation; a compliance regime that outlines consequences for failure to meet emissions targets but defers to the parties to the Protocol after it is in force to decide whether these consequences are legally binding; accounting procedures for the flexibility mechanisms; and a decision to consider at COP-8 how to achieve to a review of the adequacy of commitments that might move toward discussions of future developing country commitments.

COP-8 (New Delhi, India, 2002), COP-9 (Milan, Italy, 2003), COP-10 (Buenos Aires, 2004)

At these three meetings of the conference of parties to the UNFCCC, attempts were made to consider next steps after the 2008-2012 commitment period, but these attempts encountered resistance from developing countries and some other parties. The announced reluctance of Russia at the Milan COP-9 to undertake ratification of the Kyoto Protocol called into question whether or when the Protocol might enter into force. Without U.S. participation, the required 55% of baseline emissions of parties would not be achieved if Russia did not ratify. However, just before the COP-10 meeting, Russia did ratify the Protocol on November 18, 2004. Thus the Kyoto Protocol would enter into force 90 days later, on February 16, 2005. All three of these meetings centered on largely technical issues, and avoided major substantive declarations; what "next steps" involving developing countries should be remained a controversial issue, and was not resolved.

COP-11, Montreal, Canada

COP-11 was held in Montreal, Canada, November 28 - December 9, 2005. This was a concurrent meeting — the 11th meeting of the parties to the UNFCCC, and the first meeting of the parties (MOP-1) to the Kyoto Protocol. One of the key outcomes of the Kyoto Protocol MOP was adoption of the "Marrakech Accords," which outline what the Secretariat terms the "rule book" for the Protocol. Among other things, it formally launches emissions trading by providing rules for trading, and outlines the operational rules for the Clean Development Mechanism and Joint Implementation, both of which provide for credit to developed countries for projects to reduce emissions or augment sinks in developing or other eligible countries. Other rules adopted include how emissions are accounted for, guidelines on data systems needed, rules for a compliance system, and rules governing how absorption of carbon dioxide by agricultural soils and forests is to be measured. Discussions of "next steps" were considered under both the Protocol and the Convention. Negotiations under the Protocol were agreed upon that could lead to new binding commitments for Kyoto Protocol parties after 2012; and a decision under the Framework Convention was made to open a non-binding "dialogue on long-term cooperative action," which could include all parties to the Convention, including the United States.

Asia Pacific Partnership on Clean Development and Climate

On July 27, 2005, a six-nation partnership was announced at South East Asian Nations (ASEAN) forum. This is a partnership agreement among six nations — the United States, Australia, China, India, Japan, and South Korea. The participants described the focus of the partnership as technology development and reduction of greenhouse gas intensity, with voluntary participation. Representatives of the six nations met in Sydney, Australia, in early January, and spelled out the purpose and provided a work plan for the partnership in statements on January 12, 2006. The purposes included to "Create a voluntary, non-legally binding framework for international cooperation to facilitate the development, diffusion, deployment, and transfer of existing, emerging and longer-term cost-effective, cleaner, more efficient technologies and practices among the Partners through concrete and substantial cooperation so as to achieve practical results." Eight task forces were established to review the status of their sectors with regard to clean development and climate, and to identify cost and performance objectives and realistic goals; they focused on specific sectors: (1) Cleaner Fossil Energy, (2) Renewable Energy and Distributed Generation, (3) Power Generation and Transmission, (4) Steel, (5) Aluminum, (6) Cement, (7) Coal Mining, and (8) Buildings and Construction. The charter for the partnership states: "The partnership will be consistent with and contribute to our efforts under the UNFCCC and will complement, but not replace, the Kyoto Protocol." (More detailed information is available on the State Department website).

Congressional Interest and Activities

Discourse in Congress over the prospect of global warming and what the United States could or should do about it has yielded, over the last several years, a range of legislative proposals. Issues dealt with in bills that have been introduced in the 109th Congress include regulating emissions of carbon dioxide along with emissions of sulfur dioxide, nitrogen oxides, and mercury in so-called "multi-pollutant" legislation (see CRS Report RL32755, *Air Quality: Multi-Pollutant Legislation in the 109th Congress*); greenhouse gas reduction and carbon dioxide emissions trading systems (see CRS Report RS21581, *Climate Change: Senate Proposals to Reduce Greenhouse Gas Emissions* and CRS Report RS21067, *Global Climate Change: Controlling CO_2 Emissions Cost-Limiting Safety Valves*); energy issues relevant to climate change, especially those associated with encouraging or authorizing energy efficiency and alternative energy sources (see CRS Issue Brief IB10041, *Renewable Energy: Tax Credit, Budget, and Electricity Production Issues* and CRS Issue Brief IB10020, *Energy Efficiency: Budget, Oil Conservation, and Electricity Conservation Issues*); carbon sequestration technologies and methodologies; federal and national research concerning the prospect of abrupt climate change, climate change impacts, and climate system surprises; federal spending on climate change science programs and climate change technology programs and, more broadly, on global change research programs; and long-term research and development programs to develop new technologies to help stabilize greenhouse gas emissions. (For an overview and comparison of key climate change legislation, see CRS Report RL32955, *Climate Change Legislation in the 109th Congress*).

LEGISLATION

P.L. 109-58 (H.R. 6)

The Energy Policy Act of 2005 contains provisions addressing climate change, and a number of others were debated during consideration of the bill. Title XVI establishes programs to promote the adoption of technologies to reduce greenhouse gas intensity, similar to those adopted on the Senate floor in S.Amdt. 817. The House version of H.R. 6 did not contain climate change or greenhouse gas provisions. Two other climate change amendments included in the Senate bill as passed were not included in the final version of the bill: S.Amdt. 866 expressed the sense of the Senate that climate change is a serious concern and that Congress should establish mandatory, market-based limits on greenhouse gas emissions; and S.Amdt. 826, based largely on S. 1151, would have required mandatory emission reduction and a cap-and-trade system. H.R. 6 passed the House April 21, 2005. It passed the Senate June 28, 2005. The conference report, H.Rept. 109-190, was agreed to by the House July 28,

2005, and by the Senate July 29, 2005. The bill was signed by the President and enacted into law on August 8, 2005.

H.R. 759 (Gilchrest)/S. 342 (McCain)

To provide for a program of scientific research on abrupt climate change, to accelerate the reduction of greenhouse gas emissions in the United States by establishing a system of tradeable allowances that will limit greenhouse gas emissions in the United States. Introduced February 10, 2005. Referred to the House Committee on Science and Committee on Energy and Commerce.

H.R. 955 (Olver)

To amend the Clean Air Act to establish an inventory, registry, and information system of U.S. greenhouse gas emissions to inform the public and private sectors concerning, and encourage voluntary reductions in, greenhouse gas emissions. Introduced February 17, 2005. Referred to the Committee on Energy and Commerce; referred March 14, 2005, to Subcommittee on Energy and Air Quality.

H.R. 3057 (Kolbe)

The Foreign Operations, Export Financial and Related Programs Appropriations Act of 2006. Section 6074 would require that not less than $180 million shall be made available to support clean energy, and other climate change policies and programs in developing countries, of which $100 million is to directly promote and deploy energy conservation, efficiency, renewable and clean technologies, and the balance to measure, monitor, and reduce greenhouse gas emissions, increase carbon sequestration, and enhance climate change mitigation and adaptation programs. In addition, a climate change report is required to describe federal agency obligations and expenditures, domestic and international, for climate change in FY2005, FY2006, and FY2007 for the U.S. Agency for International Development. On June 24, 2005, the House Committee on Appropriations reported an original bill (H.Rept. 109-52). Passed the House on June 28, 2005. On June 30, 2005, the Senate Committee on Appropriations reported H.R. 3057 with an amendment in the nature of a substitute and an amendment to the title (S.Rept. 109-96). Passed the Senate on July 20, 2005. On November 2, 2005, conference committee filed a report (H.Rept. 109-265). Conference report agreed to in House November 4, 2005. Conference report agreed to in Senate November 10, 2005. Signed by President, November 14, 2005; became P.L. 109-102.

H.R. 5049 (Udall of New Mexico)

Would establish a market-based system to regulate greenhouse gas emissions and to promote advanced energy research and technology development and deployment. Introduced March 29, 2006. Referred to the Committee on Energy and Commerce, and in addition to the Committees on Science, International Relations, and Education and the Workforce, in each case for consideration of such provisions as fall within the jurisdiction of the committee concerned.

S. 245 (Collins)

To provide for the development and coordination of a comprehensive U.S. research program to facilitate understanding, assessment and prediction of human-induced and natural processes of abrupt climate change. Introduced February 1, 2005. Referred to Senate Committee on Commerce, Science, and Transportation.

S. 386 (Hagel)

To direct the Secretary of State to carry out activities that promote the adoption of technologies that reduce greenhouse gas intensity in developing countries while promoting economic development. Introduced February 15, 2005. Referred to Senate Committee on Foreign Relations.

S. 387 (Hagel)

To amend the Internal Revenue Code of 1986 to provide tax incentives for investment in greenhouse gas intensity reduction projects. Introduced February 15, 2005. Referred to the Senate Committee on Finance.

S. 388 (Hagel)

To amend the Energy Policy Act of 1992 to direct the Secretary of Energy to carry out activities that promote adoption of technologies that reduce greenhouse gas intensity and to provide credit-based financial assistance and investment protection for projects that employ advanced climate technologies or systems, and to provide for the establishment of a national greenhouse gas registry. Introduced February 15, 2005. Referred to the Senate Committee on Energy and Natural Resources.

S. 883 (Hagel)

Would direct the Secretary of State to carry out activities that promote the adoption of technologies that reduce greenhouse gas intensity in developing countries, while promoting economic development. Introduced April 21, 2005. Referred to the Committee on Foreign Relations.

S. 887 (Hagel)

Would amend the Energy Policy Act of 1992 to direct the Secretary of Energy to carry out activities that promote the adoption of technologies that reduce greenhouse gas intensity and to provide credit-based financial assistance and investment protection for projects that employ advanced climate technologies or systems, and for other purposes. Introduced April 21, 2005. Referred to the Committee on Energy and Natural Resources.

S. 745 (Byrd)

Would amend the Global Environmental Protection Assistance Act of 1989 to promote clean energy development, to open and expand clean energy markets abroad, to engage developing nations in the advancement of sustainable energy use and climate change actions, and for other purposes. Introduced April 11, 2005. Referred to the Committee on Foreign Relations.

S. 1151 (McCain)

Would provide for a program to accelerate the reduction of greenhouse gas emissions in the United States by establishing a market-driven system of greenhouse gas tradeable allowances, to limit greenhouse gas emissions in the United States and reduce dependence on foreign sources of oil, to support deployment of new climate change-related technologies, and to ensure benefits to consumers. Introduced May 26, 2005. Referred to the Committee on Environment and Public Works.

S. 2724 (Carper)

To amend the Clean Air Act (42 U.S.C. 7401 et seq.) to protect and preserve the environment and safeguard public health by ensuring that substantial emission reductions are achieved at fossil fuel-fired electric generating facilities; to significantly reduce the quantities of mercury, carbon dioxide, sulfur dioxide, and nitrogen oxides that enter the environment as a result of the combustion of fossil fuels; to achieve emission reductions from electric generating facilities in a cost-effective manner; and for other purposes. Introduced May 4, 2006. Referred to the Committee on Environment and Public Works.

S.J.Res. 5 (Feinstein)

A joint resolution expressing the sense of Congress that the United States should act to reduce greenhouse gas emission. Introduced February 16, 2005. Referred to the Committee on Foreign Relations.

In: Global Climate Change Revisited
Editor: Horace M. Karling, pp. 103-116

ISBN 978-1-59454-039-4

Chapter 6

GLOBAL CLIMATE CHANGE: THE KYOTO PROTOCOL*

Susan R. Fletcher

SUMMARY

Negotiations on the Kyoto Protocol to the United Nations Framework Convention on Climate Change (UNFCCC) were completed December 11, 1997, committing the industrialized nations to specified, legally binding reductions in emissions of six "greenhouse gases." The Protocol entered into force on February 16, 2005, and its emissions reduction requirements are binding on the 35 industrialized countries that have ratified it; the United States disengaged from the Protocol in 2001 and has not ratified it.

As structured in the negotiations completed in 1997, this treaty would commit the United States — if it were to ratify the Protocol — to a target of reducing greenhouse gases by 7% below 1990 levels during a "commitment period" between 2008-2012. Because of the fact that "sinks," which remove and store carbon from the atmosphere, are counted and because of other provisions discussed in this report, the actual reduction of emissions within the United States that would be required to meet the target was estimated to be lower than 7%.

The United States signed the Protocol on November 12, 1998. However, the Clinton Administration did not submit the Protocol to the Senate for advice and consent, acknowledging that one condition outlined by S.Res. 98, passed in mid-1997 — meaningful participation by developing countries in binding commitments limiting greenhouse gases — had not been met. In late March 2001, the Bush Administration rejected the Kyoto Protocol. The United States continued to attend the annual conferences of the parties (COPs) to the UNFCCC, but did not participate in Kyoto Protocol-related negotiations. In February, 2002, President Bush announced a U.S. policy for climate change that will rely on domestic, voluntary actions to reduce the "greenhouse gas intensity" (ratio of emissions to economic output) of the U.S. economy by 18% over the next 10 years.

* Excerpted from CRS Report RL30692 dated July 21, 2005.

Following the ratification of the Kyoto Protocol by Russia in November 2004, it entered into force on February 16, 2005, and as noted above, obligates the 35 developed nations that have ratified it to meet their commitments to reduce their greenhouse gas emissions by the amounts specified in the Protocol over the period 2008 to 2012. A total of 141 nations, including developing countries — which have no binding obligations under the treaty — had ratified the Protocol when it entered into force. In order to enter into force, the Protocol had to be ratified by developed (Annex I) nations representing 55% of the level of their emissions in 1990. Only those countries that have ratified the Protocol are bound by its terms; therefore, although the United States had signed the Protocol, it is not bound by its terms, since it has not ratified it.

This report is intended to provide background on the Kyoto Protocol and its terms. It will be updated annually or as events warrant.

BACKGROUND

Responding to concerns that human activities are increasing concentrations of "greenhouse gases" (such as carbon dioxide and methane) in the atmosphere, most nations of the world joined together in 1992 to sign the United Nations Framework Convention on Climate Change (UNFCCC). The United States was one of the first nations to ratify this treaty. It included a legally non-binding, voluntary pledge that the major industrialized/developed nations would reduce their greenhouse gas emissions to 1990 levels by the year 2000, and that all nations would undertake voluntary actions to measure, report, and limit greenhouse gas emissions.

However, as scientific consensus grew that human activities are having a discernible impact on global climate systems, significantly contribuing to a warming of the Earth that could result in major impacts such as sea level rise, changes in weather patterns, and health effects — and as it became apparent that major nations such as the United States and Japan would not meet the voluntary stabilization target by 2000 — parties to the treaty decided in 1995 that it would be necessary to enter into a legally binding, not voluntary, agreement. Negotiations began on a protocol to establish legally binding limitations or reductions in greenhouse gas emissions. It was decided by the parties that this round of negotiations would establish limitations only for the developed countries (the 38 nations listed in Annex I to the UNFCCC, including the former Communist countries, and referred to as "Annex I countries." (Developing countries are referred to as "non-Annex I countries.")[1] This was referred to as the "Berlin Mandate," which reflected continuation of the principle established in the UNFCCC that parties bore "common but differentiated responsibilities" in dealing with climate change issues, and that first steps in reducing greenhouse gas emissions should be taken by the Annex I countries.

During negotiations that preceded the December 1-11, 1997, meeting in Kyoto, Japan (COP-3), little progress was made, and the most difficult issues were not resolved until the final days — and hours — of the Conference. There was wide disparity among key players especially on three items: (1) the amount of binding reductions in greenhouse gases to be required, and the gases to be included in these requirements; (2) whether developing countries should be part of the requirements for greenhouse gas limitations; and (3) whether to include emissions trading and joint implementation, (which allow credit to be given for emissions reductions to a country that provides funding or investments in other countries that bring about the actual reductions in those other countries or locations where they may be cheaper to attain).

Following completion of the Protocol in December of 1997, decisions and details regarding a number of the more difficult issues remained to be negotiated and resolved (see below). At the fourth Conference of the Parties (COP-4) held November 2-13, 1998, in Buenos Aires, Argentina, it was apparent that these issues could not be resolved at this meeting as had been expected. Instead, parties established a two-year "Buenos Aires Plan of Action" (BAPA) to deal with these issues, with a deadline for completion at the COP-6 meeting in The Hague, Netherlands, November 13-24, 2000.

The difficulty in resolving these issues was underlined by the collapse of discussions at COP-6 in The Hague, without agreement. As discussed below, COP-6 resumed in the latter half of July 2001, following U.S. President Bush's announcement that the United States would not continue participating in negotiations on the Kyoto Protocol, and would develop an alternative approach. The U.S. delegation remained on the sidelines, declining to negotiate the key remaining issues related to the Protocol. During the July meeting, the other parties reached final agreement on most outstanding issues, with additional agreements reached on some details at the next meeting, COP-7, in Marrakech, Morocco, October 29-November 9, 2001. (See discussion below) In the annual meetings of parties to the UNFCCC in the years since the Marrakech meeting, major continuing issues regarding the Kyoto Protocol — including the issues related to future participation of developing countries — were not taken up or resolved.

In order to enter into force, the Protocol had to be ratified by nations representing 55% of these emissions. In November, Russia ratified the protocol, completing the requirements for bringing the Protocol into force within 90 days, and on February 16, 2005, it officially entered into force. The Protocol's provisions will apply only to those countries that have ratified it.

MAJOR PROVISIONS OF THE KYOTO PROTOCOL

The Kyoto Protocol was opened for signature March 16, 1998, and entered into force February 16, 2005, becoming legally binding for countries that have ratified it. The Protocol conditions required that it would be in force only when 55 nations had ratified it, provided that these ratifications included Annex I Parties that account for at least 55% of total Annex I carbon dioxide emissions in 1990.

On November 12, 1998, the United States signed the Protocol, in part because the Clinton Administration wanted to revitalize what was seen as some loss of momentum during COP-4. However, the treaty was not subsequently submitted to the Senate for approval in recognition of S.Res. 98, the 1997 resolution indicating disapproval of any treaty that did not include legally binding commitments for developing countries. In the United States, ratification of treaties can occur only after being submitted to and approved by the U.S. Senate. In 2001, early in his first term, President George W. Bush rejected the Kyoto Protocol, as noted above and discussed below, and U.S. policy has disengaged from formal negotiations on the Protocol.

The major commitments in the treaty on the most controversial issues are as follows:

Emissions Reductions

The United States would be obligated under the Protocol to a cumulative reduction in its greenhouse gas emissions of 7% below 1990 levels for three major greenhouse gases, including carbon dioxide, (and below 1995 levels for the three other, man-made gases), averaged over the commitment period 2008 to 2012. The Protocol states that Annex I Parties are committed —individually or jointly — to ensuring that their aggregate anthropogenic carbon dioxide equivalent emissions of greenhouse gases do not exceed amounts assigned to each country in Annex B to the Protocol, "with a view to reducing their overall emissions of such gases by at least 5% below 1990 levels in the commitment period 2008 to 2012." Annex A lists the 6 major greenhouse gases covered by the treaty[2].

Annex B to the Kyoto Protocol lists 39 nations, including the United States, the European Union plus the individual EU nations, Japan, and many of the former Communist nations (the same countries as Annex I to the UNFCCC). The amounts for each country are listed as percentages of the base year, 1990 (except for some former Communist countries), and range from 92% (a reduction of 8%) for most European countries — to 110% (an increase of 10%) for Iceland. The United States agreed to a commitment on this list to 93%, or a reduction of 7%below 1990 levels, to be achieved as an average over the five year commitment period, 2008-2012.

Based on projections of the growth of emissions using current technologies and processes, the reduction in greenhouse gas emissions required of the United States would likely be between 20% and 30% below where it would be otherwise by the 2008-2012 budget period.[3] However, inclusion of greenhouse gas sinks[4] — which the Protocol adopted as urged by the United States — and emissions trading, means that the domestic U.S. emission reductions from fossil fuels needed to meet a 7% target would be substantially less. However, two of the most difficult issues unresolved at Kyoto, and responsible in large part for the breakdown of the COP-6 negotiations in November 2000, are related to (1) emissions trading — specifically, how much of a country's obligation to reduce emissions can be met through purchasing credits from outside, vs. taking domestic action; and (2) the extent to which carbon sequestration by forests, soils and agricultural practices can be counted toward a country's emission reductions.

Developing Country Responsibilities

The United States had consistently taken a firm position that "meaningful participation" of developing countries in commitments made in the Protocol is critical both to achieving the goals of the treaty and to its approval by the U.S. Senate. This reflects the requirement articulated in S.Res. 98, passed in mid-1997, that the United States should not become a party to the Kyoto Protocol until developing countries are subject to binding emissions targets. The U.S. government also argued that success in dealing with the issue of climate change and global warming would require such participation. The developing country bloc argued that the Berlin Mandate — the terms of reference of the Kyoto negotiations established at COP-1 in 1995 — clearly excluded them from new commitments in this Protocol, and they continued to oppose emissions limitation commitments by non-Annex I countries.

The Kyoto Protocol was concluded without such commitments, and the Clinton Administration indicated that it would not submit the Protocol for Senate consideration until meaningful commitments were made by developing countries. At COP-4 in Buenos Aires, Argentina — host country of the meeting — became the first developing country to indicate that it will make a commitment to take on a binding emissions target for the period 2008-2012. Kazakhstan also announced its intention to take similar action. At this meeting, the United States announced it would sign the Kyoto Protocol, which it did on November 12, 1998. To date, none of the largest developing countries, such as China, India or Brazil, have shown a willingness to make commitments to reducing greenhouse gas emissions.

The Protocol does call on all Parties — developed and developing — to take a number of steps to formulate national and regional programs to improve "local emission factors," activity data, models, and national inventories of greenhouse gas emissions and sinks that remove these gases from the atmosphere. All Parties are also committed to formulate, publish, and update climate change mitigation and adaptation measures, and to cooperate in promotion and transfer of environmentally sound technologies and in scientific and technical research on the climate system.

Emissions Trading and Joint Implementation

Emissions trading, in which a Party included in Annex I "may transfer to, or acquire from, any other such Party emission reduction units resulting from projects aimed at reducing anthropogenic emissions by sources or enhancing anthropogenic removals by sinks of greenhouse gases" for the purpose of meeting its commitments under the treaty, is allowed and outlined in Article 6, with several provisos. Among the provisos is the requirement that such trading "shall be supplemental to domestic actions." The purpose of this proviso is to make it clear that a nation cannot entirely fulfill its responsibility to reduce domestic emissions by relying primarily on emissions trading or joint implementation to meet its targets. Joint implementation is project-based activity in which one country can receive emission reduction credits when it funds a project in another country where the emissions are actually reduced.

One of the more contentious issues in the negotiations concerning how the Kyoto Protocol would work has been this issue of "supplementarity" — finding agreement on what proportion of a nation's obligations could be met through these mechanisms versus domestic actions to reduce emissions within a nation's own borders. At the negotiations in Bonn, Germany, in mid-2001 (COP-6 "bis" discussed below), this issue was resolved with language that indicated there would be no quantitative limit on the credit a country could claim from use of these mechanisms, but that domestic action must constitute a significant element of the efforts of each Annex B country to meet their targets.

A number of specific issues related to the rules on how joint implementation and emissions trading would work were left at Kyoto to be negotiated and resolved in subsequent meetings; in the years since the Protocol was completed, it became increasingly clear that this is an extremely complex issue. The European Union has designed a system to be used within the EU whether or not the Kyoto Protocol enters into force, and they have set a goal of initiating emissions trading in early 2005. However, citing scientific issues yet to be resolved, the EU system will not include sinks or land use issues.

Another major "mechanism" for meeting obligations in the Protocol is provided by the establishment of a "clean development mechanism" (CDM), through which a modified form of joint implementation between developed and developing countries would occur. The United States had pushed hard for joint implementation, and early proposals were formulated with the expectation that "JI" projects would be primarily bilateral. Instead, negotiations resulted in agreement to establish the clean development mechanism to which developed Annex I countries could contribute financially, and developing/non-Annex I countries could benefit from financing for approved project activities; Annex I countries could then use certified emission reductions from such projects to contribute to their compliance with part of their emission limitation commitment. Emissions reductions achieved through this mechanism could begin in the year 2000 to count toward compliance in the first commitment period (2008-2012). Like emissions trading, making the CDM operational appears likely to be a difficult and complex process.

Buenos Aires Action Plan

Although it had been expected just after the 1997 Kyoto conference that the November, 1998, COP-4 meeting in Buenos Aires, Argentina, would resolve some of the more difficult issues left unresolved in Kyoto, it became clear during the year leading up to COP-4 that parties were far from agreement on all of these issues. Additional time for parties to analyze, negotiate, and work on these issues would be required. Therefore, the parties arrived in Buenos Aires with an agenda focused on formulating an "action plan" that would allow for the needed additional work to be done. It was decided that the work plan should be completed by the end of 2000, and should focus on the key issues, including the following:

- Rules and guidelines for the "market-based mechanisms" that allow flexibility to parties in meeting their obligations. These include emissions trading, joint implementation, and the Clean Development Mechanism (CDM). The list of critical issues to be considered include whether there should be quantified limits on how much of a country's emission reduction requirement could be met through these mechanisms, as argued by the European Union, or no quantified limit, as argued by the United States; "transparency" in making it possible to effectively track emission units; allocating risk in emissions trades — including the question of assigning liability, or responsibility, when emissions trading involves "false" credits; and key measurement, reporting and verification issues.
- Rules and procedures that would govern compliance, including provisions covering non-compliance with the treaty's commitments. This issue was left entirely open at Kyoto and remained one of the major challenges facing negotiators.
- Issues concerning development and transfer of cleaner, lower-emitting technologies, particularly to developing countries.
- Consideration of the adverse impacts of climate change and also the impacts of measures taken to respond to it, an issue of particular importance to developing

countries, who argue the need for financial assistance in order to help them cope with these impacts.

Carbon Sinks

Another issue under active negotiation and consideration by the parties, but outside the action plan itself, is defining application of the concept of carbon sinks, including how to measure and verify the categories of carbon sinks. The scientific panel that provides analysis to the parties, the Intergovernmental Panel on Climate Change (IPCC), conducted a comprehensive study on land use, land-use change, and forestry activities to identify their roles as carbon sinks and deal with the measurement and verification issues related to them.

Following the release of this report, which indicated that a large amount of carbon could be stored in a variety of carbon sinks, including not only forests, but in soils, vegetation, grazing lands, etc., the United States made a comprehensive proposal for the COP-6 negotiations to broaden the scope of acceptable carbon sinks. The Kyoto Protocol accepts in principle that a nation's forests — management practices, reforestation or afforestation — may be included in the accounting of net greenhouse gas emissions and their reduction. This would be important to the United States, as its large land area and extensive potential for greater absorption of carbon due to land management changes could greatly reduce the amount of emissions reductions needed from energy production. In a submission to the Secretariat of the UNFCCC, the United States proposed in late summer 2000 that elaboration at COP-6 of land use changes acceptable under the Protocol should also include soil carbon sequestration and vegetation.

Few decisions were reached, nor were they expected, on the more difficult issues outlined in the Buenos Aires Plan of Action at the COP-5 meeting in Bonn, Germany, held October 25-November 24, 1999.

COP-6 Negotiations, the Hague

The November 2000 COP-6 meeting in The Hague was the last negotiation session on the Kyoto Protocol in which the United States participated. After President George W. Bush became President in January 2001, he rejected the Kyoto Protocol and opted out of participation in Kyoto-related negotiations.

The negotiations at the COP-6 meeting were illustrative of some of the most contentious issues of the Kyoto Protocol, and the difficulties of reaching agreement on them led to the ultimate collapse of these talks. Leading up to the COP-6 meeting, there were two intersessional negotiation sessions of the UNFCCC Subsidiary Bodies (the key mechanisms for considering Kyoto issues between COP meetings) in 2000, June 12-16 and September 11-15, dealing with the issues of the Buenos Aires workplan and attempting to fashion a negotiating text for final consideration at the November 13-24 COP-6 meeting. At the conclusion of the meeting in Lyon, it was reported that language on aspects of some issues had been agreed upon, but that the negotiating text had been expanded to some 200 pages, much of it "bracketed." (Brackets indicate that no agreement has been found on the language in brackets, and often several alternative possibilities are reflected within brackets.) Thus, though there was negotiating text on most issues, disagreements remained on most key issues. Observers reported that political positions remained entrenched and little movement toward compromise appeared evident in Lyon. The end result was that many participants felt doubtful that many of the key issues could be completely resolved in November, 2000, at COP-6.

As talks began at COP-6 in The Hague, Netherlands, on November 13, they centered initially on the "Buenos Aires Plan of Action" (BAPA), but evolved into a high-level negotiation over the major political issues. These included major controversy over the United States' proposal to allow credit for carbon "sinks" in forests and agricultural lands, satisfying a major proportion of the U.S. emissions reductions in this way (between half and one-quarter, according to various versions of the U.S. proposal); disagreements over consequences for non-compliance by countries that did not meet their emission reduction targets; and difficulties in resolving how developing countries could obtain financial assistance to deal with adverse effects of climate change and meet their obligations to plan for measuring and possibly reducing greenhouse gas emissions.

In the final hours of COP-6, despite some compromises agreed between the United States and some EU countries, the EU countries as a whole, reportedly led by Denmark and Germany, rejected the compromise positions, and the talks in The Hague collapsed. Jan Pronk, the President of the COP, suspended COP-6 without agreement. Discussions between the EU and the "Umbrella group" that includes the United States, Canada, Japan and Australia, were held in Ottawa, Canada, during the week of December 4, 2000, in order to try and salvage some of the agreement reached at the end of the talks in The Hague. However, the U.S. negotiators reported that these talks were "inconclusive" and the differences were still in place, or even exacerbated, after this meeting.

At the end of 2000, based on discussions at The Hague and in Ottawa, the issues particularly in contention were as follows:

- *Mechanisms, especially emissions trading*: The main issue here was "supplementarity" — the position of the United States was that there should not be quantitative limits to the amount of emissions reductions that are allowed toward a country's obligations through emissions trading or joint implementation. The EU and others argue there should be such limitations, in order to force nations to take more extensive domestic action to reduce emissions. This issue is related to the commitment outlined in the Kyoto Protocol that emissions trading should be "supplemental" to domestic action. The United States was supported by the "umbrella group" in which it is joined by New Zealand, Japan, Canada, Australia, Russia, Ukraine, Norway and Iceland. Another related issue was whether carbon sinks can be included in the Clean Development Mechanism (CDM) in which a contributing developed country can claim credit for actions to reduce emissions in developing countries. There were significant divisions on this issue not among developed countries, but among developing countries, as well.
- *Compliance issues:* Decisions on how non-compliance with Protocol (and UNFCCC) obligations should be handled was a very controversial issue. The United States position was that there should be binding consequences, but these should be in the form of additional obligations in subsequent commitment periods, and not in the

form of financial penalties. There was considerable disagreement at The Hague over whether financial penalties should be allowed. Agreement on binding consequences would probably require an amendment to the Protocol, which would be separately agreed to, and separately ratified. Opponents to binding consequences include Japan, Russia, and Australia, who were concerned, among other things, about opening the Protocol to an amendment process. On the structure of a compliance regime, there was substantial agreement on a likely outcome: a single compliance body with two functions or branches — (1) to facilitate and assist compliance (mainly for developing countries), and (2) an enforcement function where decisions would be made on whether compliance violations have occurred and what consequences should be applied. Consequences under consideration, in addition to financial penalties, included losing access to mechanisms like emissions trading and/or subtracting from future allocations of allowable carbon emissions.

- *Land use and land use change and forestry (LULUCF):* As noted above, the Kyoto Protocol accepts in principle that a nation's forests —management practices, reforestation or afforestation — may be included in the accounting of net greenhouse gas emissions and their reduction. The United States proposed at COP-6 that land use changes acceptable under the Protocol should also include soil carbon sequestration and vegetation. Major issues were how to attain precision in measuring absorption and release of carbon from land-based sources, how permanent such land use mechanisms would be, and the extent to which land use absorption counted by a nation is "additional" to business as usual.

  The United States made a series of controversial proposals on how to count carbon sequestration, beginning with basically counting most of the carbon sequestration in its extensive forest cover toward its obligations. It subsequently revised this proposal and put forward a formula that included three parts: a first "interval" allowing up to 20 million tons of carbon to be counted at 100% for any country with forests absorbing that much; a second interval of a certain amount in which credit for a certain percentage would be allowed up to a certain threshold; then full credit for tons absorbed beyond the threshold (which would be historically determined in relation to baseline absorption amounts). The EU opposed the U.S. proposal, mainly on the issue of forests, and the extent to which a country like the United States would receive credits for a "business as usual" scenario that did not involve the harder emissions reductions from fuel sources and technological measures. When the United States put numbers to this proposal, the U.S. credits from carbon sinks appeared to represent about 125 million tons of carbon, against a likely need to reduce emissions by about 600 million tons of carbon to meet its commitment in 2008-2012. This was strongly opposed by the EU and other countries, and a stalemate over this issue, despite several revisions downward of the U.S. position and tentative acceptance of a much smaller amount by the EU, was thought to be a major factor in the collapse of the November COP-6 negotiations at The Hague.
- *Developing country participation:* The United States had been seeking additional commitments from developing countries in a series of informal discussions and consultations during the period since the Kyoto Protocol was completed in late 1997, but only Argentina and Kazakhstan have shown a willingness to make such

commitments. Little willingness to do so has been shown by other developing countries. The possibility that it would be discussed under an agenda item dealing with "adequacy of commitments" did not occur at The Hague.

Bush Administration Policy

Just over two months after the COP-6 talks collapsed, President George W. Bush took office, and announced his Administration would carry out a cabinet-level review of climate policy. However, the Bush Administration announced in late March 2001 that it would not be interested in continuing discussion on the Kyoto Protocol, which was characterized as "dead" in terms of U.S. policy. The cabinet-level review of U.S. climate policy was on-going, and the Administration indicated that it would be interested in pursuing alternative approaches or cooperative efforts such as market-based incentives and voluntary measures, to address climate change concerns.

The EU nations and others, such as Japan, expressed deep concern and dismay at this new U.S. position, and a high-level delegation of EU officials visited the United States to attempt to re-engage the United States in the Kyoto Protocol process. This effort was reported as having been "rebuffed." Some observers began considering the implications of the new U.S. position for the Protocol, and whether the EU and other nations would ratify the Protocol, and try to bring it into force without the United States.

President Bush made a policy statement in mid-June, 2001, resulting from the continuing cabinet-level review of climate change options, in which he confirmed the U.S. approach as rejecting the Kyoto Protocol and favoring voluntary actions, increased scientific research, and market mechanisms. President Bush also outlined a U.S. Climate Change Research Initiative, and the National Climate Change technology Initiative. During his mid-June trip to Europe, President Bush discussed climate policy with European heads of state and met strong opposition to the U.S. position. The outcome was that Europe and the United States would "agree to disagree" on climate. The Europeans announced they would proceed with ratification of the Kyoto Protocol, without the United States if necessary.

President Bush indicated that the United States would continue to participate in international meetings on climate change, but would not negotiate on Kyoto Protocol issues. When the COP-6 meeting resumed in Bonn, Germany, in July 2001, the U.S. delegation followed this approach, declining to participate in negotiations on most Kyoto Protocol issues, and remaining on the sidelines as observers. The other parties reached agreement on the key political issues and announced that the developed countries would move toward ratification of the Kyoto Protocol without the United States.

COP-6 Resumed, Bonn, Germany

When the COP-6 negotiations resumed July 16-27, 2001, in Bonn, Germany, little progress had been made on resolving the differences that had produced an impasse in The Hague. However, this meeting took place after President George Bush had rejected the Kyoto Protocol in March; as a result, the United States delegation to this meeting declined to participate in the negotiations related to the Protocol and chose to act as observers at that meeting. As the other parties negotiated the key issues, agreement was reached on most of the major political issues, to the surprise of most observers given the low level of expectations that preceded the meeting. The agreements included:

(1) Mechanisms — the "flexibility" mechanisms which the United States had strongly favored as the Protocol was initially put together, including emissions trading; joint implementation; and the Clean Development Mechanism (CDM), which provides funding from developed countries for emissions reduction activities in developing countries, with credit for the donor countries. One of the key elements of this agreement was that there would be no quantitative limit on the credit a country could claim from use of these mechanisms, but that domestic action must constitute a significant element of the efforts of each Annex B country to meet its targets.
(2) Carbon sinks — credit was agreed to for broad activities that absorb carbon from the atmosphere or store it, including forest and cropland management, and revegetation, with no over-all cap on the amount of credit that a country could claim for sinks activities. In the case of forest management, an Appendix Z establishes country-specific caps for each Annex I country; for example, a cap of 13 million tons could be credited to Japan (which represents about 4% of its base-year emissions). For cropland management, countries could receive credit only for carbon sequestration increases above 1990 levels.
(3) Compliance — final action on compliance procedures and mechanisms that would address non-compliance with Protocol provisions was deferred to COP-7, but included broad outlines of consequences for failing to meet emissions targets that would include a requirement to "make up" shortfalls at 1.3 tons to 1, suspension of the right to sell credits for surplus emissions reductions; and a required compliance action plan for those not meeting their targets.
(4) Financing — three new funds were agreed upon to provide assistance for needs associated with climate change; a least-developed-country fund to support National Adaptation Programs of Action; and a Kyoto Protocol adaptation fund supported by a CDM levy and voluntary contributions.

A number of operational details attendant upon these decisions remained to be negotiated and agreed upon, and these were the major issues of the COP-7 meeting that followed.

COP-7 and Subsequent Conferences of the Parties

At the COP-7 meeting in Marrakech, Morocco October 29-November 10, 2001, negotiators in effect completed the work of the Buenos Aires Plan of Action, finalizing most of the operational details and setting the stage for nations to ratify the Protocol. The United States delegation continued to act as observers, declining to participate in active negotiations. Other parties continued to express their hope that the United States would re-engage in the process at some point, but indicated their intention to seek ratification of the requisite number of countries to bring the Protocol into force (55 countries representing 55% of developed country emissions of carbon dioxide in 1990).

The main decisions at COP-7 included operational rules for international emissions trading among parties to the Protocol and for the CDM and joint implementation; a compliance regime that outlines consequences for failure to meet emissions targets but defers to the parties to the Protocol after it is in force to decide whether these consequences are legally binding; accounting procedures for the flexibility mechanisms; and a decision to consider at COP-8 how to achieve to a review of the adequacy of commitments that might move toward discussions of future developing country commitments. Other parties reiterated their hope that the United States would re-engage with the international cooperation efforts under the Protocol, but also their decision to seek ratification by their governments and the Protocol's entry into force by the WSSD in 2002, as noted above.

At COP-8 (New Delhi, 2002), COP-9 (Milan, 2003), and COP-10 (Buenos Aires, 2004) over the subsequent three years, no major policy issues were negotiated or decided. The next Conference of the Parties, COP-11, is scheduled to take place in Montreal, Canada, November 28 - December 9, 2005. This will be concurrent meetings of the 11th Conference of the Parties to the UNFCCC and the first Meeting of the Parties to the Kyoto Protocol; it is expected that discussion at this meeting will focus on key issues involved in "next steps," including what might be outlined as steps to be taken by developing countries. It appears likely that this discussion will be controversial among the parties, and that a key consideration will be the question of how greenhouse gas emissions can be addressed without posing obstacles for economic development in developing countries.

CURRENT U.S. POLICY

Disengagement from the Kyoto Protocol remains U.S. policy, with an emphasis instead on domestic action. On February 14, 2002, President Bush announced a U.S. policy for climate change, based on voluntary domestic measures, characterized as a "new approach for meeting the long-term challenge of climate change." The centerpiece of this announcement was the plan to reduce greenhouse gas intensity of the U.S. economy by 18% over the next 10 years. Greenhouse gas intensity measures the ratio of greenhouse gas emissions to economic output and has been declining in the United States over the past several years. The Administration stated that the goal, to be met through voluntary action, is to achieve efficiency improvements that would reduce the 183 metric tons of emissions per million dollars of gross domestic product (GDP) to 151 in 2012. The plan notes that "if, in 2012, we find that we are not on track toward meeting our goal, and sound science justifies further policy action, the United States will respond with additional measures that may include a broad, market-based program" and other incentives and voluntary measures to accelerate technology development.

In addition, the plan directs the Secretary of Energy in consultation with other key agencies, to "substantially improve the emission reduction registry" to upgrade the voluntary emission reduction program under section 1605(b) of the 1992 Energy Policy Act, to bring about enhanced measurement accuracy, reliability, and verifiability. Other measures include providing for protected, transferable emission reduction credits, increased funding of $700 million in total climate-related spending, and a new management structure to coordinate climate change and technology research. Domestic policies such as tax incentives for renewable energy and new technology, development of fuel-efficient vehicles and cleaner fuels, and carbon sequestration were also proposed, along with several international bilateral initiatives and relatively modest increases in foreign assistance.

Some observers praised the plan for taking a practical, conservative approach to government action and for relying on voluntary measures. Critics observed that voluntary approaches by themselves have not historically often been effective and noted that the reductions in energy intensity are very little different from current trends and would allow for significant increases in over-all greenhouse gas emissions rather than reductions.

Ratification Issues

For the United States to ratify the Kyoto Protocol, the treaty would have to be transmitted to the U.S. Senate by the President for advice and consent. A two-thirds majority vote in the Senate is required for approval. As long as the United States has not ratified the treaty, it is not subject to its terms and obligations. President Clinton expressed strong support for the Kyoto Protocol, though criticizing it for not including commitments for developing countries. During his administration, the United States signed the Protocol on November 12, 1998. The U.S. signature was criticized by several Members of Congress who opposed the treaty on a number of grounds, including questions about the scientific justification for it and about the likely economic impacts that might occur if the United States were to attempt to meet its emission reduction commitments in the treaty.

In recognition of the opposition expressed in the Senate by S.Res. 98, which passed 95-0 in the months before the 1997 Kyoto meeting, to a Protocol that does not include requirements for emissions limitations by developing countries, President Clinton did not submit the treaty to the Senate for advice and consent, citing lack of meaningful developing country participation. Given the position announced by the Bush Administration, opposing the Kyoto Protocol, it continues to be unlikely that it will be transmitted to the Senate.

Legislation

A number of bills have been introduced in the 109th Congress to address ways of reducing greenhouse gas emissions in the United States. Debate during consideration of the omnibus energy bill, H.R. 6, in the Senate included votes on several amendments, including some that were adopted on technology development, deployment and transfer, and a Sense of the Senate resolution that recognizes that human activities are a substantial cause of greenhouse gas accumulations in the atmosphere, causing average temperatures to rise. This resolution also states that "Congress should enact a comprehensive and effective national program of mandatory market based limits and incentives on emissions of greenhouse gases that slow, top, and reverse the growth of such emissions..." in a way that avoids harm to the U.S. economy and encourages comparable action by other countries. (For additional information, see CRS Report RL32955, *Climate Change Legislation in the 109th Congress*).

ENDNOTES

[1] For additional information on the negotiations in Kyoto and related background, see CRS Issue Brief IB89005, *Global Climate Change*.

[2] The six gases covered by the Protocol are carbon dioxide (CO_2), methane (CH_4), nitrous oxide (N_2O), hydrofluorocarbons (HFCs), perfluorocarbons (PFCs), and sulphur hexafluoride (SF_6). The most prominent of these, and the most pervasive in human economic activity is carbon dioxide, produced when wood or fossil fuels such as oil, coal, and gas are burned.

[3] See CRS Report 98-235 ENR, *Reducing Greenhouse Gases: How Much from What Baseline?*

[4] Greenhouse gases, especially CO_2, are absorbed by a number of processes in forests, soils, and other ecosystems. These are called "sinks."

In: Global Climate Change Revisited
Editor: Horace M. Karling, pp. 117-130
ISBN 978-1-59454-039-4

Chapter 7

CLIMATE CHANGE: THE KYOTO PROTOCOL AND INTERNATIONAL ACTIONS*

Susan R. Fletcher

SUMMARY

The first treaty to address climate change, the United Nations Framework Convention on Climate Change (UNFCCC), was completed and opened for signature in 1992. This treaty includes commitments to establish national action plans for voluntary measures that could reduce greenhouse gas emissions to 1990 levels in order to begin mitigating possible global warming. The United States was one of the first nations to sign and ratify this treaty, and it entered into force in 1994.

However, it was soon concluded by parties to the treaty that mandatory reductions in the six major greenhouse gases (of which carbon dioxide, mainly from burning of fossil fuels, is the most prevalent) would be required. The resulting Kyoto Protocol, which was completed in 1997 and entered into force in February 2005, committed industrialized nations that ratify it to specified, legally binding reductions in emissions of the six major greenhouse gases. The United States has not ratified the Protocol, and thus is not bound by its provisions. One provision of the Protocol that has attracted particular interest among business interests in nations that are parties to it is for "emissions trading" and other "flexibility" mechanisms in which nations can contribute to meeting their reductions by purchasing "credits" from other countries that have reduced their emissions beyond their requirements.

The United States signed the Protocol in late 1998, but President Clinton did not submit the Protocol to the Senate for approval because conditions outlined by S.Res. 98, passed in mid-1997 — including meaningful participation by developing countries in binding commitments limiting greenhouse gases — had not been met. In March 2001, the Bush Administration rejected the Kyoto Protocol, and subsequently announced a U.S. policy for climate change that relies on voluntary actions to reduce the "greenhouse gas

* Excerpted from CRS Report RL33826 dated January 24, 2007.

intensity" (ratio of emissions to economic output) of the U.S. economy by 18% over the next 10 years.

Each of the industrialized nations listed in Annex B of the Kyoto Protocol has a specified emissions target. Overall, the collective commitments are to reduce the Parties' emissions by at least 5% below their 1990 levels, averaged over the "commitment period" 2008 to 2012. On a "parallel track" of activities, the United States took an initiative in 2005, the Asia-Pacific Partnership for Clean Development and Climate, together with China, India, Japan, Australia, and South Korea, which is a voluntary effort (though without specific targets) to reduce the greenhouse gas intensity of their economies through technology cooperation.

As of December 2006, the UNFCCC Secretariat reported that 168 nations and the European Union have ratified or accepted the Kyoto Protocol. Annual meetings of the parties continue, and attention during the negotiations has turned in large part to "next steps" following the end of the commitment period in 2012, as well as a review of the effectiveness of the Protocol. Major challenges remain to find agreement on the nature of commitments, if any, that would prove acceptable to all major players: current parties, developing countries that are major emitters, and the United States.

Introduction and Overview

Responding to concerns that human activities are increasing concentrations of "greenhouse gases" (such as carbon dioxide and methane) in the atmosphere and causing potentially damaging climate change, nearly all nations of the world joined together in 1992 to sign the United Nations Framework Convention on Climate Change (UNFCCC). The United States was one of the first nations to ratify this treaty. It included a legally non-binding, voluntary pledge that the major industrialized/developed nations would establish national action plans aiming to reduce their greenhouse gas emissions to 1990 levels by the year 2000, and that all nations would undertake voluntary actions to measure and report greenhouse gas emissions to the UNFCCC Secretariat.

The parties to the UNFCCC hold annual meetings called Conferences of the Parties (COPs), at which unresolved issues are negotiated, rules of procedure are established or amended, and reviews of progress are considered. As scientific consensus grew that human activities are having a discernible impact on global climate systems, contributing to a warming of the Earth that could result in major impacts such as sea level rise, changes in weather patterns, and health effects — and as it became apparent that many major nations such as the United States and Japan would not be able to reduce their emissions to 1990 levels by 2000 — parties to the treaty decided in 1995 that it would be necessary to move beyond voluntary measures and to enter into legally binding commitments. Negotiations began on a protocol to establish legally binding limitations or reductions in greenhouse gas emissions. It was decided by the parties that this round of negotiations would, in keeping with principles established in the UNFCCC, establish limitations only for the developed countries — the 38 nations listed in Annex I to the UNFCCC, including the former Communist countries, plus the European Union, and referred to as "Annex I countries." Developing countries are referred to as "non-Annex I countries."

A basic principle established in the UNFCCC and continuing in negotiations on the Kyoto Protocol is that parties have "common but differentiated responsibilities" in dealing with climate change issues, and that first steps in reducing greenhouse gas emissions should be taken by the Annex I countries. Because developed countries have emitted the largest share of the greenhouse gases already in the atmosphere (carbon dioxide releases remain in the atmosphere for many decades), and because they are wealthier and more able to incur costs of any necessary changes in their economies, it was agreed by negotiators of the UNFCCC, and subsequently the Kyoto Protocol, that this principle would be a basic tenet of climate negotiations.

The Kyoto Protocol negotiations were completed in late 1997. The protocol establishes legally binding, mandatory emissions reductions for the six major greenhouse gases1. It requires that Annex I countries (listed again in Annex B of the Protocol) reduce their aggregate greenhouse gas emissions by 5% below 1990 levels (1990 is the baseline for carbon dioxide, methane, and nitrous oxide; 1995 is the baseline year for the other 3 gases), averaged over the "commitment period" of 2008 to 2012. Each country was assigned individually negotiated targets which differed according to their situations.

The United States signed the Protocol in 1998. However, no country is subject to the provisions of a treaty until it has been ratified, which in the United States requires the consent of the U.S. Senate. Because the Senate was on record in mid-1997 in S.Res. 98, objecting to a treaty that had no mandatory obligations for developing countries, President Clinton did not submit the Protocol to the Senate for advice and consent; therefore the United States did not ratify the Protocol during his Administration. Then in March 2001, soon after President George W. Bush took office, he rejected the Kyoto Protocol, and the United States declined further participation in Kyoto Protocol negotiations. After several years, the required number of Annex I countries had ratified (all but the United States and Australia), and the Protocol entered into force in February 2005. As of December 2006, 168 nations plus the European Union had ratified the Protocol, representing 61.6% of Annex I countries' 1990 greenhouse gas emissions. Since the United States has not ratified the Protocol, it is not subject to its terms.

Negotiations continued after the 1997 treaty was finalized in order to put in place the detailed rules for how the Kyoto Protocol would operate and to establish procedures for how its provisions would be carried out. These negotiations were continued through two subsidiary bodies that address technical issues, and then decisions were made at the annual conferences of the parties (COPs) to the UNFCCC until February 2005 when the Kyoto Protocol had achieved the necessary ratifications to enter into force; the annual meeting is now a UNFCCC COP, combined with a "meeting of the parties (MOP)" to the Protocol. Thus the annual meetings are now referred to as COP/MOP meetings. The United States continues to participate in the discussions and negotiations of the COPs, but as it is not a party to the Kyoto Protocol, it does not participate in Kyoto-related negotiations.

KEY PROTOCOL PROVISIONS

The major commitments in the treaty on the key issues are as follows:

Obligations of All Parties

The Kyoto Protocol calls on all Parties — developed and developing — to take a number of steps outlined in Article 10 to contribute to scientific research and monitoring of the climate system and greenhouse gases in their countries. They are also committed to formulate national and regional programs to improve local emission factors; carry out steps to promote and transfer environmentally sound technologies; strengthen national capacity building activities; and conduct national inventories of greenhouse gas emissions and sinks that remove these gases from the atmosphere.

In keeping with the principle of common but differentiated responsibilities, the UNFCCC and the Kyoto Protocol recognize the relatively low per capita greenhouse gas emissions in developing countries, and the need to take into account the development activities of these countries that will require increased energy use. The Protocol does not impose any binding requirements on developing countries; the commitments in Article 10 are regarded as essentially voluntary.

Obligations to reduce greenhouse gas emissions by Annex I countries are the focus of most of the Kyoto Protocol, which outlines the legally binding emissions reductions that the Parties to the treaty are to undertake, and provides for development of procedures and rules that apply to Parties as they move toward meeting these binding obligations. These are briefly summarized below.

Emissions Reductions

The Kyoto Protocol states that Annex I Parties are committed — individually or jointly — to ensuring that their aggregate anthropogenic carbon dioxide equivalent emissions of greenhouse gases do not exceed amounts assigned to each country in Annex B, "with a view to reducing their overall emissions of such gases by at least 5% below 1990 levels in the commitment period 2008 to 2012." Negotiations on the Kyoto Protocol included a nation-by-nation allocation of the percentage each Annex I country would be obligated to reduce its greenhouse gas emissions in order to collectively reach the overall 5% reduction agreed to in the Protocol.

Annex B to the Kyoto Protocol lists 39 nations, including the United States, the European Union plus the individual EU nations, Japan, and many of the former Communist nations (the same countries as Annex I to the UNFCCC). The amounts for each country are listed in this annex as percentages of the base year, 1990 (except for some former Communist countries, which use a more recent year), and range from 92% (a reduction of 8%) for most European countries — to 110% (an increase of 10%) for Iceland. In negotiations on the Protocol, the United States agreed to a commitment on this list to 93%, or a reduction of 7% below 1990 levels. These commitments refer to averages that would be below each Party's 1990 levels for three major greenhouse gases, including carbon dioxide, (and below 1995 levels for the three other, man-made gases), averaged over the "commitment period" 2008 to 2012. (As noted above, only nations that ratify the Protocol are subject to its terms; the United States later rejected participation, and thus is not bound by it.)2

However, two of the most difficult issues unresolved in 1997 at Kyoto and still under discussion are related to counting emissions of a nation, specifically how to take into account:

(1) emissions trading — specifically, how much of a country's obligation to reduce emissions can be met through purchasing credits from outside, vs. taking domestic action; and (2) the extent to which carbon sequestration by forests, soils and agricultural practices can be counted toward a country's emission reductions. These are discussed below.

Implementation: "Flexibility" Mechanisms

Emissions Trading

Emissions trading is one of three "flexibility" mechanisms contained in the Kyoto Protocol (article 17).[3] Under the Kyoto Protocol, developed countries are given greenhouse gas emissions "budgets" (or emissions "caps") for the compliance period 2008-2012 based on a percentage of their 1990 or 1995 emissions levels (depending on the particular greenhouse gas). If a country determined that it would exceed its emissions limit during the compliance period, emissions trading would permit it to purchase emissions reductions "credits"[4] from another country that determined it would achieve more emissions reductions than necessary to comply. With emissions trading, countries that can make relatively inexpensive emissions reductions have an incentive to reduce emissions below the level required by the Kyoto Protocol, and sell the extra credits to other countries whose emissions control costs are more expensive. Thus, both the seller and the buyer would have lower costs by virtue of the seller's profit and the buyer's savings. This type of implementation scheme is commonly called a "cap-and-trade" program.[5]

This mechanism, however, comes with significant restrictions under the Kyoto Protocol. First, emissions trading is restricted to countries that have legally binding greenhouse gas emission limitations — the Annex 1 parties, which as noted above includes only developed, industrialized countries that have ratified the Protocol.[6] Another requirement is that emissions credits must "be supplemental to domestic actions for the purpose of meeting quantified emission limitations and reduction commitments...."[7] However, the Protocol is vague as to what "supplemental" means, and the term is subject to continuing interpretation.

Currently, the largest emissions trading scheme in use under the Kyoto Protocol is the European Union's Emissions Trading system (ETS). The EU-ETS is a cornerstone of the EU's efforts to meet its obligation under the Kyoto Protocol, and it currently covers more than 11,500 energy-intensive facilities across the now 27 EU Member countries, including oil refineries, powerplants over 20 megawatts in capacity, coke ovens, and iron and steel plants, along with cement, glass, lime, brick, ceramics, and pulp and paper installations. Covered entities emit about 45% of the EU's carbon dioxide emissions. The trading program does not cover emissions of greenhouse gases, which account for about 20% of the EU's total non-CO_2 greenhouse gas emissions. The first trading period began January 1, 2005. A second trading period is scheduled to begin in 2008, covering the period of the Kyoto Protocol, with a third period planned for 2013.[8]

Clean Development Mechanism (CDM)

The Kyoto Protocol supplements the cap-and-trade implementation scheme discussed above with two project-based schemes that permit Annex 1 countries to obtain additional credits that they can use to meet their emission caps. The first is CDM — the only mechanism under the Kyoto Protocol that involves non-Annex 1 countries. Under its provisions, industrialized countries can receive Certified Emissions Reduction credits (CERs) for reductions achieved from a greenhouse gas reduction project in a "host" non-Annex 1 country. CERs can then be used by the industrialized country to meet its compliance requirements. The process is overseen by a CDM Executive Board that registers and validates projects, issues CERs, and manages a series of panels and working groups. A critical component of the process is the requirement that CERs issued under the CDM represent only reductions in excess of those that would have occurred in the absence of the project. CERs can be issued from appropriate projects initiated after 2000. In 2005, China, India, and Brazil were estimated by Point Carbon to be responsible for about 63% of the total volume of all projects at the Project Design Document (PDD) stage of development.9 In the case of China, large volumes of CDM contracts are primarily the result of a few large HFC-23 reduction projects.

Joint Implementation (JI)

In contrast, JI is a program in which industrialized countries can receive Emission Reduction Units (ERUs) from greenhouse gas reduction projects conducted jointly between two Annex 1 countries. Like CERs, ERUs can be used by the participating countries for compliance purposes. There are two tracks under JI (called Track 1 and Track 2). Track 2 mirrors the process used by the CDM but involves different institutions. Track 1 is a simplified process that puts more of the responsibility on the host country. Like the CDM, ERUs issued under JI must represent reductions achieved in addition to those that would have occurred in the absence of the project. Unlike CDM projects, ERUs can only be transferred beginning in 2008. Romania has been the most active host JI country. However, it should be noted that in 2005, the volumes involved in the project markets were overwhelmingly the result of CDM projects, which accounted for 93% of tonnage transacted (397 million metric tons (MMt) compared with 28 MMt for JI).[10]

Carbon "Sinks"

One of the most contentious issues in the negotiations over Kyoto Protocol rules has been how to give nations credit for carbon "sinks": forests and land uses that absorb (sequester) carbon from the atmosphere and have the effect of reducing the net additions a country makes to atmospheric CO2 levels. This has been negotiated under the term "Land Use, Land Use Change and Forestry" (LULUCF). Issues include how to allocate credit for existing forest cover (of which some nations, like the United States, have a great deal and others have very little) and what actions in relation to LULUCF would constitute legitimate carbon reductions. Only increased sequestration above 1990 levels, achieved by specific sequestration activities, would be counted.

The final decisions were that only certain activities would be eligible for use as offsets against Protocol obligations: afforestation (planting forest cover where there had been none); reforestation (re-planting tree cover where it had been removed); deforestation prevention; forest management; cropland management; grazing land management; and revegetation. The rules of the Protocol do not put an overall cap on sinks for countries, but instead incorporated country-specific limits on each of the categories of sinks activities, listed in an Appendix Z.

Exactly how carbon absorbed in sinks will count toward a nation's obligations is still the subject of on-going discussion and refinement.

Compliance Mechanism

Achieving agreement among Kyoto Protocol negotiators on compliance — in particular, penalties for non-compliance by Annex B Protocol parties — was difficult, and took several years after the Protocol was finalized in 1997. This was one of several controversial issues that was resolved as part of the Marrakech accords at the COP-7 meeting in 2001. In the final rules on compliance, parties agreed that if a Party falls short of its emissions target during the first commitment period (2008 to 2012), it must make up the difference in the second commitment period plus a penalty of 30%. Such a party will lose its eligibility for emissions trading, and must develop a "compliance action plan."

This decision included establishment of a Compliance Committee, composed of a plenary, an operational bureau, and two branches: the Facilitative Branch and the Enforcement Branch. The Facilitative Branch is intended to provide advice and assistance to Parties, and to provide an "early warning" to Parties that may be in danger of not complying; the Enforcement Branch would have the responsibility of applying consequences for Parties that do not meet their commitments. A Protocol rulebook provides procedures for considering cases of non-compliance or possible non-compliance, and a procedure for reviewing the cases regarding eligibility to participate in the Protocol's financial and other mechanisms.

Compliance involves not only meeting emissions reductions commitments, but also preparation of adequate GHG inventories and several other procedural requirements, for which there are no penalties. However, the mechanism's penalties for failure to meet emissions reductions targets would come into play only when the commitment period is well underway; there remain a number of uncertainties as to how it will function. It does appear, as discussed below, that many parties to the Protocol may find that achieving their emissions reductions obligations will prove to be difficult or impossible within the commitment period.

Prospects for Compliance by Kyoto Protocol Parties

As parties to the Kyoto Protocol contemplate future commitments in the post-2012 period, often called "post-Kyoto commitments," the question of whether existing commitments are likely to be met is important. This section reviews the status of the parties to the Protocol in relation to their existing commitment and current GHG emissions. Many are facing major challenges to achieving emissions reductions, given the increase in emissions they have had in recent years.

Status of Annex I Countries on Compliance

The Kyoto Protocol mandates compliance over a five-year averaging period —2008 to 2012. Thus, it is an uncertain business attempting to forecast what countries will be achieving throughout the entire compliance period. For example, in September 2005, Point Carbon did an assessment of compliance efforts and policies of all countries that are expected to fall short of their Kyoto targets. As noted by Point Carbon, the assessment was a snapshot of the situation — countries are refining their policies continuously and, in the case of the EU, compliance will be determined on an EU-15 basis (the EU bubble).[11] As indicated in Figure 1, the assessment does indicate that several countries are challenged in meeting their Kyoto commitments. In particular, countries in southern Europe, such as Italy and Spain, are forecasted to have emissions substantially above the targets assigned to them under the EU bubble. In contrast, countries like Sweden and the United Kingdom are anticipated to achieve or maintain lower than required emission levels during the compliance period.

For the EU, the situation puts considerable pressure on the European Commission's (EC) current review of Member countries' National Allocation Plans for the Kyoto commitment period. The EU has consistently stated that it will meet its commitments under the Kyoto Protocol and is currently developing targets for the post-2012 period.

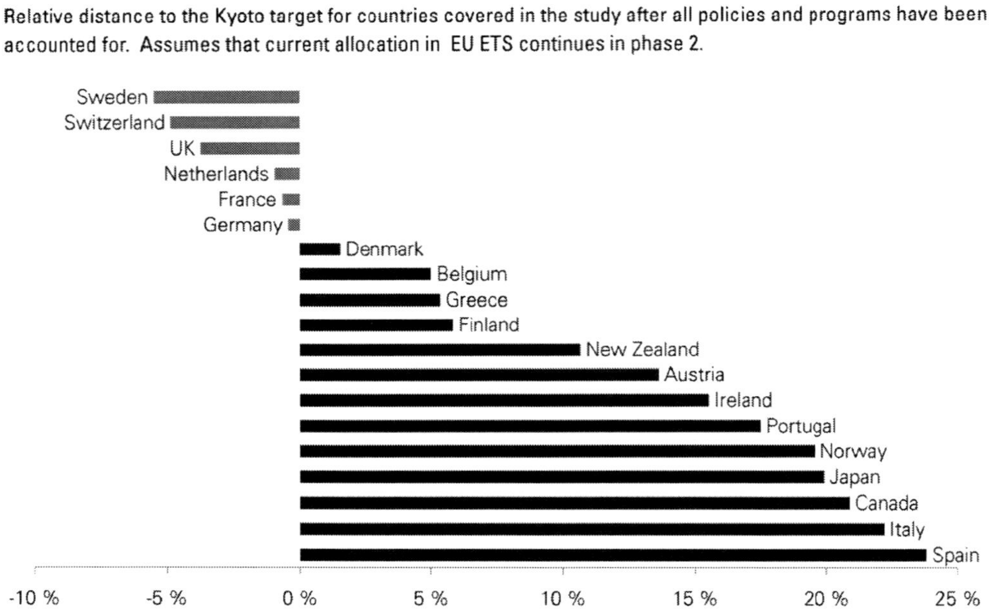

Source: Point Carbon, *Carbon 2006: Towards a Truly Global Market* (February 2006), p. 7.

Figure 1. September 2005 Point Carbon Assessment of Compliance Efforts.

Because of the flexibility mechanisms discussed above, potential compliance with Kyoto is not necessarily threatened by emissions growth in some countries, since such nations can purchase emissions credits to offset some of their emissions increases. Indeed, current analyses of credits available from economies in transition (mostly eastern European countries) indicate that sufficient surplus credits will exist in such countries to cover the deficit created by emissions growth in other countries. Specifically, as indicated by Figure 2, surplus credits are anticipated in Russia, Ukraine, and other eastern European countries — sufficient surplus credits for other countries anticipated to have deficits (such as parts of the

EU, Japan, Canada, and New Zealand) to purchase them in order to cover their compliance requirements.

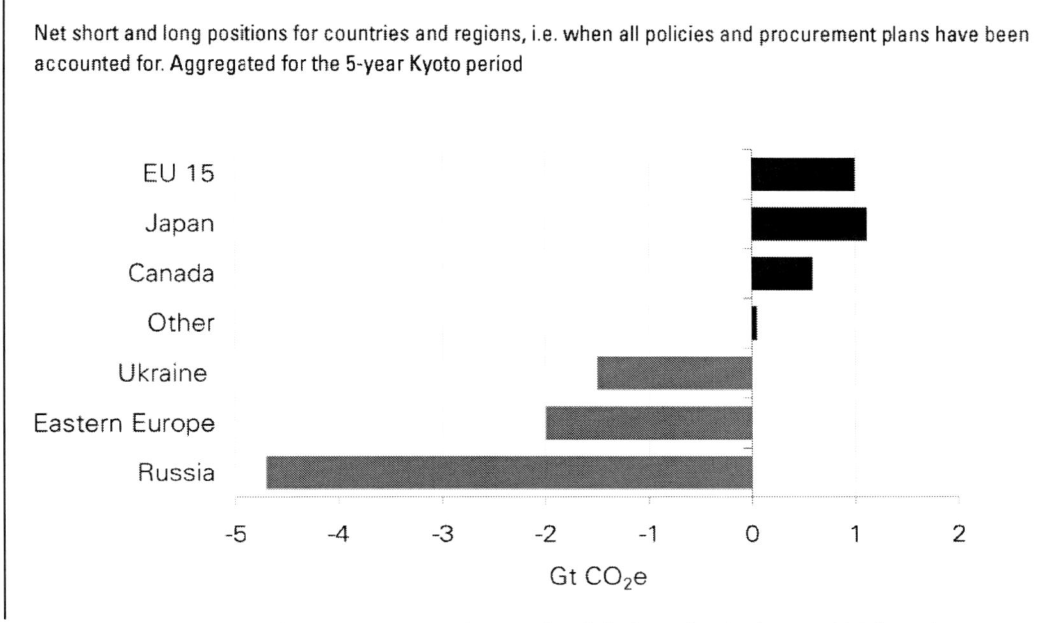

Source: Point Carbon, *Carbon 2006: Towards a Truly Global Market* (February 2006), p. 8.

Figure 2. 2005 Point Carbon Assessment of Potential Credit Supply.

This long position with respect to credits does not include anticipated CERs and ERUs from CDM and JI projects. As indicated by Figure 3, the potential for additional credits from these sources is substantial, particularly from China. Of course, there is a range of risk with respect to these sources (including the possibility that any given project may not be built or may not be certified as valid for issuing CERs or ERUs), but indications are that additional credits will be available from these sources.

There are some potential potholes on the road to compliance. Perhaps the most important is the requirement that credits gained through emissions trading and joint implementation be "supplemental to domestic actions." This requirement under Article 17 has been subject to extensive discussion and interpretation. Resisting efforts by the EU to place a quantitative limit on the procurement of credit via these mechanisms, the Conference of Parties decided to put no quantitative limit on such mechanisms, but that "the use of the mechanism shall be supplemental to domestic action and domestic action shall thus constitute a significant element of the effort made by each party."[12] No means of determining the significance of domestic actions has been finalized. The EU as a whole has taken a strict interpretation of "supplementarity" (although some individual countries have indicated they would like to use the mechanism more extensively), while other countries, such as Japan, are anticipated to use the mechanisms substantially in meeting their requirements.

Another major issue to be resolved is the volatility of the credit markets at the current time. Credit prices for the ETS have ranged from less than 5 euro a ton to about 30 euro a ton

— currently the price is less than 5 euro. Such volatility makes planning, implementation, and participation in emissions reduction programs difficult.

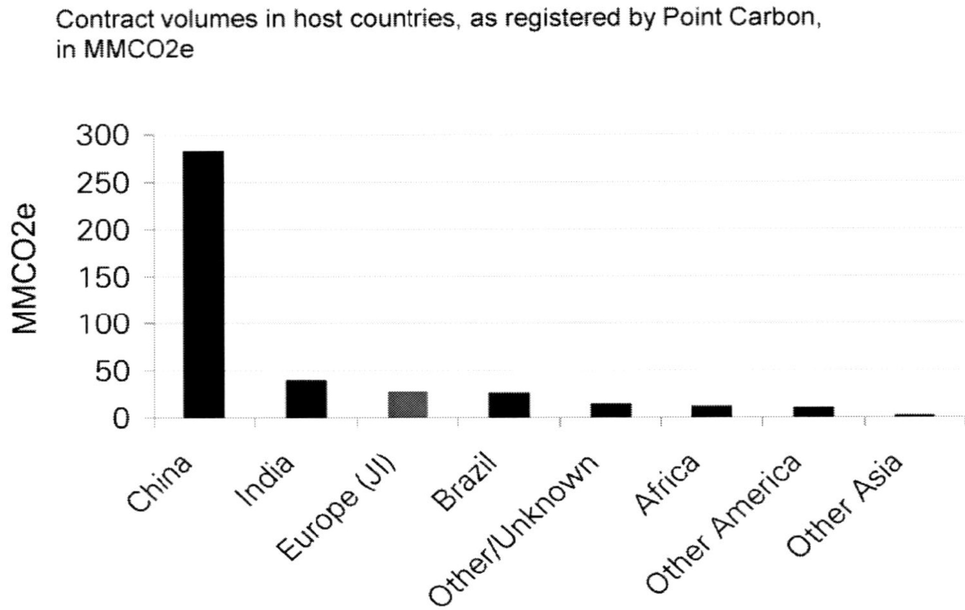

Source: Point Carbon, *Carbon 2006: Towards a Truly Global Market* (February 28, 2006) p. 23.

Figure 3. 2005 Point Carbon Assessment of CDM and JI Transactions.

U.S. Position

The United States took an active role in the negotiations that shaped the 1997 Kyoto Protocol. It signed the Protocol in 1998, but President Clinton did not submit it to the Senate for approval, citing the Byrd-Hagel resolution, S.Res. 98, that passed unanimously just prior to the completion of the Protocol in Kyoto. This non-binding resolution expressed the sense of the Senate that it would not support a treaty that did not include obligations for developing countries, or that would harm the U.S. economy.

In March 2001, President George W. Bush rejected the Kyoto Protocol, and the United States has not formally participated in further negotiations on the Protocol. Until 2005, meetings to negotiate details and procedures of the Kyoto Protocol's rules occurred annually in conjunction with the conference of parties (COP) to the UNFCCC. The United States attended these meetings and participated in discussions or negotiations pertaining only to the UNFCCC; it also has been active behind the scenes in discussions of Kyoto Protocol issues, but has not formally participated. Regaining U.S. participation and ratification of the Kyoto Protocol remains a major goal for parties to the Kyoto Protocol, along with engaging major developing country participation, since meaningful reduction in global greenhouse gas emissions will depend on reductions in emissions by the largest emitters.

In February 2002, President Bush announced a U.S. policy of reducing the net "greenhouse gas intensity" of the U.S. economy by 18% over the next 10 years.[13] Greenhouse gas intensity measures the ratio of greenhouse gas emissions to economic output. GHG intensity has been declining in the United States, and the 18% reduction goal compares to a projected "business as usual" decline in intensity of 14%[14] for the 10-year period. According to some, the 18% goal would not achieve reductions of greenhouse gases significantly more than the existing trend.

The Administration stated that the goal, to be met through voluntary action, is to achieve efficiency improvements that would reduce the 183 metric tons of emissions per million dollars of gross domestic product (GDP) expected under "business as usual" to 151 metric tons in 2012. The plan noted that "if, in 2012, we find that we are not on track toward meeting our goal, and sound science justifies further policy action, the United States will respond with additional measures that may include a broad, market-based program" and other incentives and voluntary measures to accelerate technology development.

U.S. actions related to climate change include extensive scientific and research programs[15], and continuing international activities to cooperate on a bilateral basis or through other international forums to support voluntary action. One of the most prominent international U.S. activities recently was the initiative taken in the Asia-Pacific Partnership on Clean Development and Climate, discussed below.

Asia-Pacific Partnership on Clean Development and Climate

On July 27, 2005, the United States announced formation of a six-nation Asia-Pacific Partnership on Clean Development and Climate (commonly denoted by the acronym 'APP'), which would work on cooperatively reducing greenhouse gas intensity of their economies. This partnership agreement included three of the world's largest emitters of greenhouse gases: the United States, China, and India, plus Japan, Australia, and South Korea. Of the three developed country nations in the partnership — the United States, Australia, and Japan — only Japan has ratified the Kyoto Protocol.

The participants described the focus of the partnership as technology development and reduction of greenhouse gas intensity, on a voluntary basis. No specific targets were announced. The members of the partnership indicated that their goal was to "complement, but not replace the Kyoto Protocol." However, concerns have been raised by some observers as to whether this partnership and its approach would in fact reinforce the resistance of the two largest developing country greenhouse gas emitters — China and India — to taking on mandatory emissions commitments in the next phase of the Kyoto Protocol.

The purposes identified by partnership members include to "Create a voluntary, non-legally binding framework for international cooperation to facilitate the development, diffusion, deployment, and transfer of existing, emerging and longer-term cost-effective, cleaner, more efficient technologies and practices among the Partners through concrete and substantial cooperation so as to achieve practical results."[16]

Task forces focused on eight industrial or business sectors were established to review the status of their sectors with regard to clean development and climate, to identify cost and performance objectives and realistic goals, and report on recommended actions within their sectors. The eight Task Forces are: (1) Cleaner Fossil Energy, (2) Renewable Energy and

Distributed Generation, (3) Power Generation and Transmission, (4) Steel, (5) Aluminum, (6) Cement, (7) Coal Mining, and (8) Buildings and Construction. The task force reports, termed Action Plans, are now complete and can be accessed on the partnership website.

The Executive Summary of the Task Force Action Plans describes the projects to be undertaken, which include sectoral assessments, identifying best practices, capacity building, and technology research and demonstration. These activities are described as "a significant first step toward a more comprehensive set of actions to address clean development and climate goals.[17]

The remaining questions include the degree to which funding and investment will be forthcoming for the work plan goals. Also unclear is the degree to which the developing country partners will undertake activities in addition to those receiving outside funding. Another question — critical to the future of the Kyoto Protocol, and raised initially when the partnership was first announced — is whether the two largest developing country emitters of greenhouse gases, China and India, will see this parallel effort as indeed complementary to the Kyoto Protocol, or whether they will continue to resist binding emissions reductions in the next commitment period of the Kyoto Prtocol in favor of the non-binding, voluntary approach espoused by the Asia-Pacific Partnership, with greenhouse gas intensity reduction as their focus.

Focus of Future Negotiations: Next Steps and Post-Kyoto Commitments

The Kyoto Protocol was always intended to be a first step in moving toward reducing global accumulations of greenhouse gases in the atmosphere. Negotiators recognized that the goals of the Protocol, even if met by all the parties, would not produce the stabilization of atmospheric greenhouse gases posited as the goal of the UNFCCC. The Protocol set forth a timetable for reviewing progress of actions undertaken to meet the Protocol's goals and to consider "next steps." It has been generally anticipated that next steps after 2012 would consider measures to be taken by both developed and developing countries. Throughout the process developing countries have been unwilling to make binding commitments on greenhouse gas limitations or management.

At the first "Meeting of the Parties" of the Kyoto Protocol in November/December 2005 (the 11th COP of the UNFCCC) — COP-11/MOP-1 —in Montreal, Canada, both the United States and developing countries were resistant to the idea of negotiating new legally binding commitments for the post-Kyoto (after 2012) period.[18] The Kyoto Protocol parties were also reluctant to discuss new commitments in the post-Kyoto period if they did not include all the major emitters — including the United States, China and India. By the end of that meeting a compromise was reached in which two processes were set in motion to consider next steps:

- An "Ad hoc Working Group" (AWG) was established under the Protocol to begin consideration of next steps for developed country parties in the Post-Kyoto period; since the United States is not a party to the Protocol, it does not play a role in this process (except, of course, as an observer — from which vantage point its position has generally been made known and generally taken into account by the parties).

- A non-binding, two-year "dialogue on long-term cooperation" was launched under the auspices of the UNFCCC that includes the United States and all parties to the UNFCCC (virtually all of the world's countries). The decision on establishing the dialogue specifically ruled out including any negotiations leading to new commitments. Its goals are to support implementation of existing commitments under the Convention, support voluntary actions by developing countries, and to support development of national and international responses to climate change. The areas of focus for these discussions are: sustainable development, adaptation, technology development and transfer, and market-based opportunities.

Both of these processes involve workshops and meetings with reports to the COP/MOP meetings in 2006 and 2007. Neither involves any deadlines for completion of the discussions or negotiations.

At COP-12/MOP-2 held in Nairobi, Kenya, in November 2006, both the AWG and the Dialogue were involved in workshops and discussions, and the AWG formulated a work program for future meetings that involve analytical subjects that would underlie any consideration of post-2012 targets for developed countries. The Nairobi meeting also included the beginning of a review of the Protocol's effectiveness — a somewhat controversial issue because of the implications the review process is thought to have for future commitments. Little progress was made at this COP/MOP, however, and decisions on the scope and content of the review were put off until the COP/MOP 4 to be held in 2008. Issues related to adaptation to climate change focused primarily on administration of an Adaptation Fund established to assist adaptation efforts in developing countries.

The Kyoto Protocol commitment period begins in 2008, and it was widely expected when the Protocol was negotiated that by that time, next steps for the post-2012 period would be either decided or under active negotiation. However, the challenge remains to find agreement on the nature of commitments, if any, that would be acceptable to major players — including Kyoto Protocol parties with existing obligations, developing countries that are major emitters, and the United States, whose role is regarded as critical by all potential participants in the post-2012 period. These issues will continue to be considered at COP-13/MOP-3 in early December 2007, to be held in Bali, Indonesia.

ENDNOTES

1 The six gases covered by the Protocol are carbon dioxide (CO_2), methane (CH_4), nitrous oxide (N_2O), hydrofluorocarbons (HFCs), perfluorocarbons (PFCs), and sulphur hexafluoride (SF_6). The most prominent of these, and the most pervasive in human economic activity, is carbon dioxide, produced when wood or fossil fuels such as oil, coal, and gas are burned. Concentrations of carbon dioxide in the atmosphere have increased from 280 parts per million (ppm) in 1850 to some 380 ppm in 2006.

2 When the United States signed the Protocol, and based on projections of the growth of emissions using current technologies and processes, the reduction in greenhouse gas emissions required of the United States would likely have been between 20% and 30% below where it would be otherwise by the 2008-2012 commitment period. However,

inclusion of greenhouse gas sinks [Greenhouse gases, especially CO_2, are absorbed by a number of processes in forests, soils, and other ecosystems. These are called "sinks."] — which the Protocol adopted as urged by the United States — and emissions trading, means that the domestic U.S. emission reductions from fossil fuels needed to meet a 7% target would have been substantially less.

[3] The other mechanisms are Joint Implementation (Article 6) and the Clean Development Mechanism (Article 12).

[4] A credit would generally represent the reduction of one metric ton of carbon dioxide equivalent emissions.

[5] For a detailed discussion, see CRS Report RL33799, *Climate Change: Design Approaches for a Greenhouse Gas Reduction Program*, by Larry Parker.

[6] Although called "Annex 1" countries in reference to Annex 1 of the Framework Convention on Climate Change (FCCC), the correct reference is to Annex B of the Kyoto Protocol. The lists of countries in Annex 1 and Annex B are very similar, but not identical. CRS uses the common usage term, Annex 1, in this report.

[7] Article 17, Kyoto Protocol.

[8] For more information, see CRS Report RL33581, *Climate Change: The European Union's Emission Trading System (EU-ETS)*, by Larry Parker.

[9] Point Carbon, *Carbon 2006: Towards a Truly Global Market* (26 February 2006), p. 14. Point Carbon is a company that provides in-depth analysis, forecasting, market intelligence and news about carbon emission markets.

[10] Point Carbon, *Carbon 2006: Towards a Truly Global Market* (26 February 2006), p. 22.

[11] Permitted under the Kyoto Protocol, a "bubble" is a regulatory device that permits two or more countries to be treated as one for purposes of the Protocol — their emissions can be averaged in order to reach their targets.

[12] *Preparations for the first session of the Conference of the Parties serving as the Meeting of the Parties to the Kyoto Protocol (Decision 8/CP.4): Decision 5/CP.6 implementation of the Buenos Aires Plan of Action*, FCCC/CP/2001/1.7 (July 24, 2001).

[13] See the White House website for a summary and discussion of this policy, at [http://www.whitehouse.gov/news/releases/2002/02/climatechange.html].

[14] See CRS Report 98-235, *Global Climate Change: U.S. Greenhouse Gas Emissions — Status, Trends, and Projections*, by John Blodgett and Larry Parker.

[15] See CRS Report RL33817, *Climate Change: Federal Expenditures*, by Jane A. Leggett.

[16] Extensive information about the partnership is available on its website at [http://www.asiapacificpartnership.org].

[17] See Executive Summary at [http://www.asiapacificpartnership.org]. This Summary provides examples of projects identified by each Task Force's action plans.

[18] For more detailed summaries of the COP and COP/MOP meetings, see the reports of the Earth Negotiations Bulletin (ENB) at [http://www.iisd.ca/process/climate_atm.htm] and the Pew Center on Climate Change at [http://www.pewclimate.org/what_s_being_done/in_the_world/reports/cops.cfm].

In: Global Climate Change Revisited
Editor: Horace M. Karling, pp. 131-146
ISBN 978-1-59454-039-4

Chapter 8

CLIMATE CHANGE LEGISLATION IN THE 109TH CONGRESS*

Brent D. Yacobucci

SUMMARY

Climate change and greenhouse gas (GHG) emissions are an issue in the 109th Congress, as they have been in past Congresses. Bills directly addressing climate change issues range from those focused primarily on climate change research to comprehensive emissions cap-and-trade programs for the six greenhouse gases covered under the United Nations Framework Convention on Climate Change. Additional bills focus on GHG reporting and registries, or on power plant emissions of carbon dioxide, as part of wider controls on pollutant emissions.

Within several broad categories, the bills vary in their approaches to climate change issues. For example, some bills covering research issues focus solely on modeling the effects of future climate change, whereas others address the development of monitoring technologies. Bills focusing on technology deployment do so through tax incentives and credit-based programs within the United States or by promoting deployment in developing countries. Bills with greenhouse gas registries may be voluntary or mandatory and vary in the entities covered and the gases registered. Bills with emission reduction requirements also vary in the entities covered, the gases limited, and the target emissions levels.

Most notably, on August 8, 2005, President Bush signed the Energy Policy Act of 2005 (P.L. 109-58, H.R. 6). Among other provisions, Title XVI of the bill establishes programs to promote the development and deployment of technologies to reduce greenhouse gas intensity.

This report briefly discusses the basic concepts on which these bills are based and compares major provisions of the bills in each of the following categories: climate change research, technology deployment, GHG reporting and registries, and emissions reduction programs. This report will be updated as events warrant.

---

* Excerpted from CRS Report RL32955 dated August 4, 2006.

Climate change is viewed as a global issue, but proposed responses generally require action at the national level. In 1992, the United States ratified the United Nations' Framework Convention on Climate Change (UNFCCC), which called on industrialized countries to take the lead in making voluntary efforts to reduce greenhouse gases.[1] Over the past decade, a variety of voluntary and regulatory actions have been proposed or undertaken in the United States, including monitoring of utility carbon dioxide emissions, improved appliance efficiency, and incentives for developing renewable energy sources. In 2001, President George W. Bush rejected the Kyoto Protocol to the UNFCCC, which called for legally binding commitments by developed countries to reduce their greenhouse gas emissions. Instead, the Bush Administration has focused on reducing the greenhouse gas intensity[2] of the U.S. economy. In the meantime, some states and local governments, as well as private entities, have taken actions to reduce emissions and limit the potential impacts of climate change. In light of these actions, a number of bills have been introduced in Congress to address climate change.

In the 109th Congress, numerous bills have been introduced that directly or indirectly address climate change. Several bills address the climate change issue directly, either through emissions limits, incentives for reductions, or research and information gathering on climate change and greenhouse gas emissions mitigation. This report describes and compares bills that directly address climate change, as opposed to those that address other issues but could have ancillary impacts (e.g., energy efficiency and conservation). Topics covered by these bills fall into four major categories: (1) those that would promote research on the effects of climate change and on methods to measure and predict climate change; (2) those that would create incentives for the deployment of emission-reducing technologies in the United States or other countries; (3) those that would establish greenhouse gas (GHG) monitoring systems as a basis for research or for any potential reduction program; and (4) those that would establish market-based programs to directly limit greenhouse gas emissions. These categories are not mutually exclusive, and several bills address more than one of the above categories. The major provisions of these bills are categorized in Appendix 1 and summarized in Appendix 2.

In several cases, bill sponsors have introduced modified versions of their climate change bills. For the purposes of the discussion below, it is assumed that the newest version supersedes earlier versions. These bills include S. 1151 for S. 342 (McCain); S. 883 for S. 386 (Hagel); S. 887 for S. 388 (Hagel); and S. 1203 for S. 387 (Hagel).

Energy Bill Amendments

On August 8, 2005, President Bush signed the Energy Policy Act of 2005 (P.L. 109-58, H.R. 6). Title XVI establishes a voluntary national program designed to accelerate demonstration and deployment of less-carbon-intensive technology to encourage voluntary reductions in greenhouse gases. The title attempts to support actions focused on reducing U.S. carbon intensity (the ratio of greenhouse gas emissions per unit of gross domestic product). The program would not establish a requirement to reduce emissions. This title also establishes a program to encourage exports of carbon intensity-reducing technologies to developing countries.

As part of the Senate debate over H.R. 6, several amendments on climate change were offered. S.Amdt. 817, which inserted a new Title XVI in the bill, incorporates language from S. 883 and S. 887. This amendment was agreed to on a 66-29 vote. These provisions are similar to those included in the final version of the bill. The House version of H.R. 6 did not expressly address climate change issues.

Not included in the final bill was Section 1612 of the Senate bill, which expressed the Sense of the Senate that human activities are a substantial cause of greenhouse gas accumulation in the atmosphere, causing average temperatures to rise. Further, the resolution stated that "Congress should enact a comprehensive and effective national program of mandatory market-based limits and incentives on emissions of greenhouse gases that slow, stop, and reverse the growth of such emissions at a rate and in a manner that — (1) will not significantly harm the United States economy; and (2) will encourage comparable action by other nations that are major trading partners and key contributors to global emissions." This was the first Sense of the Senate resolution on climate change since S.Res. 98 in 1997, which voiced concern over the economic effects of emissions limits and the sense that developing countries must participate in meaningful action to control emissions. S.J.Res. 5 similarly urges U.S. action on climate change. This resolution has been referred to committee.

The Senate also debated whether to adopt S.Amdt. 826, which contained language similar to S. 1151. This amendment would have established a mandatory cap-and-trade system to limit greenhouse gas emissions from covered entities to year 2000 levels by 2010. This amendment was rejected on a 38-60 vote.

Climate Change Research Bills

Global climate change is a complex issue. While most scientists agree that the climate is changing in response to greenhouse gas (GHG) emissions, uncertainties concerning the causes and the effects of climate change remain and are a continuing subject of extensive scientific research.[3] Further, research is ongoing into technologies to improve efficiency, reduce fossil fuel consumption, and sequester carbon dioxide emissions.

Research Bills

One bill in the 109th Congress, S. 245 (Collins), focuses solely on climate change research.[4] S. 245 calls for the development and testing of climate change models based on historic climatic changes, and to incorporate nonlinear aspects of geophysical systems that could lead to abrupt changes in climate.

Research Provisions in Broader Bills

In addition to S. 245, several bills include climate change research provisions as part of a broader climate change legislation. Specifically, research in S. 1151, S. 3698, H.R. 759, and H.R. 2828 would focus on abrupt climate change research and new climate change measurement technologies. H.R. 5049 would establish a new Advanced Research Projects Agency within the Department of Energy.

DEPLOYMENT OF GREENHOUSE GAS REDUCTION TECHNOLOGY

In the 109th Congress, several bills would promote the deployment and diffusion of technologies to reduce greenhouse gas emissions, either as part of broader legislation to limit greenhouse gases, or as stand-alone legislation. Deployment strategies include tax incentives for investment in technologies to improve efficiency and/or lower emissions and grants, loans, and other incentives for technology transfer to developing countries. S. 1203 (Hagel) would establish tax incentives for investment in technologies to reduce greenhouse gas intensity. S. 745 (Byrd), S. 883 (Hagel), S. 1151 (McCain), and S. 3698 (Jeffords) would establish grant and loan programs to deploy technologies in developing countries that have been developed or demonstrated in the United States. S. 887 (Hagel) would establish a credit-based deployment program for technologies to reduce greenhouse gas intensity; support would include direct loans, loan guarantees, lines of credit, and production incentive payments. The final version of H.R. 6 incorporates language similar to S. 883 and S. 887. H.R. 2828 (Inslee) provides a wide array of incentives for improvements in energy efficiency and other strategies that would reduce greenhouse gas emissions.

In addition to the above bills on technology deployment, §585(b) of the FY2006 Foreign Operations Appropriation Act (P.L. 109-102) requires the President to submit (within 60 days of submitting the FY2007 budget request) a report on federal agency expenditures (foreign and domestic) on climate change activities. The act specifically requires a report on U.S. Agency for International Development funding for climate change activities in developing countries, including technology deployment.

GHG REPORTING AND REGISTRY BILLS

Under the UNFCCC, the United States annually publishes reports on its GHG emissions.[5] The United States Environmental Protection Agency (EPA) does this reporting using various techniques (e.g., fuel analysis for CO_2). The three dominant sources of GHG emissions are electricity generation (33.1%), transportation (26.9%), and industry (19%).[6] At the national level, electric utilities must report their GHG emissions pursuant to the 1990 Clean Air Act, but there is no overall national GHG reporting requirement. However, some states also gather data through voluntary or mandatory GHG emissions reporting mechanisms.[7]

H.R. 955 (Olver) focuses primarily on expanding emissions reporting to include a broad array of sources. All entities that emit more than 10,000 metric tons of carbon dioxide equivalent must report their emissions, except that farms are exempt. Further, manufacturers and importers of automobiles and Department of Energy-listed products[8] must report the emissions from their products. The purpose of the bill is to promote greenhouse gas reductions and to generate accurate emissions data that can be used by public and private entities for various purposes.

In addition to H.R. 955, which focuses solely on GHG reporting, several other bills would require emissions monitoring and reporting as part of a program to reduce emissions of carbon dioxide or of all greenhouse gases. These emissions reductions efforts are discussed in the following section. S. 150 (Jeffords), S. 730 (Leahy), H.R. 1451 (Waxman), and H.R. 1873 (Bass) would require electricity producers to report their carbon dioxide emissions in order to determine compliance with carbon dioxide caps. S. 1151 (McCain), H.R. 759 (Gilchrest), and H.R. 2828 (Inslee) would require major emitters of all six greenhouse gases to report their emissions; the bills require reporting from entities that emit more than 10,000 metric tons (11,000 tons) of carbon dioxide equivalent. H.R. 5049 (Udall, T.) would require all fossil fuel suppliers to report the amount of greenhouse gases that will result from combustion of the fuel supplied.

GHG Emission-Reduction Bills

The United States has no federal GHG reduction requirements, though there are proposals to require such reductions. These proposals include "command and control" regulations on emissions, GHG emission taxes, and market-based techniques to limit emissions. The latter, market-based programs typically take as their model the Clean Air Act's acid rain program.[9]

In the 109th Congress, bills have been introduced that would establish market-based caps on GHG emissions. These bills are compared in Table 1. Three bills, S. 1151 (McCain), H.R. 759 (Gilchrest), and H.R. 2828 (Inslee), would cap the emissions of the six greenhouse gases specified in the United Nations' Framework Convention on Climate Change. Five other bills, S. 150 (Jeffords), S. 730 (Leahy), S. 2724 (Carper), H.R. 1451 (Waxman), and H.R. 1873 (Bass), would focus on reducing carbon dioxide from electric utilities. Each of these bills would use market-based trading mechanisms to limit GHG emissions. Cap-and-trade programs set strict limits on specific emissions from a particular group of sources, allowing individual sources to trade reductions. This flexibility in who makes reductions can lead to lower costs. In an efficient market, entities that face relatively low emission-reduction costs could achieve extra emission reductions. These entities could then sell their unused allowances to entities that face higher emission-reduction costs. An entity facing higher costs could purchase allowances that would allow it to emit more than its initial emissions allotment would otherwise permit. It should be noted that in all cases, total U.S. emissions may decrease or increase depending on the entities covered, the greenhouse gases controlled, and the emissions trading schemes.

Another market-based option is to require tradeable emissions permits, but establish a "safety valve" price. In this scenario, if the market value of a permit exceeds a set price — the safety valve — covered entities can purchase an unlimited number of permits from the government. In this way, the overall price to covered entities — and the economy — is limited, but specific emission reduction targets may not be reached. H.R. 5049 (Udall, T.) would establish a such a system.

Carbon Dioxide Reduction Bills

As shown in Table 1, S. 150, S. 730, H.R. 1451, and H.R. 1873 focus on electric utility emissions. These "multi-pollutant" bills would limit emissions of carbon dioxide, along with other air pollutants.[10] (See Table 1.) In all four cases, carbon dioxide emissions limitations would start in 2010.[11]

Comprehensive GHG Emissions Reductions

Unlike other bills proposed in the 109th Congress, the Climate Stewardship Act of 2005 (H.R. 759), the New Apollo Energy Act of 2005 (H.R. 2828), and the Climate Stewardship and Innovation Act of 2005 (S. 1151) focus on achieving market-driven reductions in all six greenhouse gases (see Table 1). The legislation applies to entities in the electricity, transportation, industry, and commercial sectors that emit over 10,000 metric tons (11,000 tons) of greenhouse gases per year. Starting in 2010, the bills would cap total GHG emissions from all these sources at 6.5 billion tons (CO_2 equivalent emissions), reduced by the amount of CO_2 (equivalent emissions) from non-covered entities in the year 2000. The bills would also establish a formula for allocating GHG emissions allowances, and a climate change credit corporation to manage allowance trading. Language similar to S. 1151 was offered as an amendment on the Senate floor to H.R. 6. This amendment was rejected on a 38-60 vote.

In addition to establishing caps on all six greenhouse gases, the above bills would support climate change research and establish a GHG emissions inventory (see above). The bills also include a requirement that the Administrator of the EPA establish a national GHG database, and develop methods and standards to measure and verify GHG emissions.

The Safe Climate Act of 2006 (H.R. 5642) and the Global Warming Pollution Act (S. 3698) would grant EPA broad authority to establish regulations such that total greenhouse gas emissions are reduced to 80% below 1990 levels by 2050. These bills do not designate covered entities or required reduction levels for specific sectors, but leave those decisions to EPA's discretion.

Safety Valve Bills

The Keep America Competitive Global Warming Policy Act of 2006 (H.R. 5049) would establish a system of allowances for fossil fuel suppliers and a safety valve of $25 per ton of carbon, indexed to inflation.

Table 1. Market-Based Greenhouse Gas Emission Caps

	S. 150 (Jeffords)	**S. 1151 (McCain), H.R. 759 (Gilchrest), H.R. 2828 (Inslee)**	**S. 730 (Leahy)**	**H.R. 1451 (Waxman)**	**H.R. 1873 (Bass)**	**S. 2724 (Carper)**
Covered sources	Any fossil fuel-fired electric generating facility that has a capacity of greater than 15 megawatts, generates electricity for sale, and emits a covered pollutant into the air.	Any electric power, industrial, or commercial entity that emits over 10,000 metric tons of CO_2 equivalent/year; any refiner or importer of petroleum products for transportation use that when combusted will emit over 10,000 metric tons of CO_2 equivalent/year; and, any importer or producer of HFCs, PFCs or SF6 that, when used, will emit over 10,000 metric tons of CO_2 equivalent/year.	All electricity generating facilities in the United States.	Any fossil fuel-fired electric generating facility that has a capacity of greater than 15 megawatts and generates electricity for sale.	Any fossil fuel-fired electric generating facility that has a capacity of greater than 25 megawatts and generates electricity for sale.	Any fossil fuel- fired electric generating facility that has a capacity of greater than 25 megawatts and generates electricity for sale.
Covered pollutants	One GHG: carbon dioxide; other Pollutants: sulfur dioxide, nitrogen oxides, and mercury.	All six GHGs.	One GHG: CO_2; other pollutants: sulfur dioxide, nitrogen oxides, and mercury.	One GHG: CO_2; other pollutants: sulfur dioxide, nitrogen oxides, and mercury.	One GHG: CO_2; other pollutants: sulfur dioxide, nitrogen oxides, and mercury.	One GHG: CO_2; other pollutants: sulfur dioxide, nitrogen oxides, and mercury.

Table 1. (Continued)

	S. 150 (Jeffords)	S. 1151 (McCain), H.R. 759 (Gilchrest), H.R. 2828 (Inslee)	S. 730 (Leahy)	H.R. 1451 (Waxman)	H.R. 1873 (Bass)	S. 2724 (Carper)
Emissions cap	Utility CO_2 emissions limited to 2.05 billion tons per year[a] beginning in 2010.	6.5 billion tons of CO_2 equivalent per year beginning in 2010 for all covered entities taken together.	Utility CO_2 emissions limited to 2.05 billion tons per year beginning in 2010.	Utility CO_2 emission cap estimated at 1.94 billion tons per year beginning in 2010.	Estimated at 2.46 billion tons in 2010, declining to 2.38 billion tons in 2015.	Estimated at 2.65 billion tons in 2010, declining to 2.45 billion tons in 2015.
Implementation Strategy	Tradeable allowance system. Allowances allocated to various sectors and interests, including households, dislocated workers and communities, electricity-intensive industries, affected utilities, energy efficiency and renewable energy activities, and sequestration activities.	Tradeable allowance system. EPA is directed to determine allocations based on several economic and equity criteria, including efficiency and impact on consumers. Allowances are to be allocated upstream to refiners and importers of transportation fuel, along with	Absolute caps on mercury emissions, no trading permitted between facilities at different sites. Implementation strategy for other pollutants to be determined by EPA.	To be determined by EPA — market mechanisms permitted (except for mercury).	Tradeable allowance system for all pollutants; allocations based on historic electricity output. CO_2 program includes allowance allocations for incremental nuclear capacity and renewable energy.	Tradeable allowance system varies by pollutant. For CO_2, allocations based on historic electricity output. CO_2 program includes allowance allocations for incremental nuclear capacity and renewable energy, along with sequestration and early action provisions.

		producers of HFCs, PFCs, and SF_6; downstream to electric generation, industrial, and commercial entities.				
Percentage change in CO_2 emissions v. business as usual by 2010[b]	-7.5%	-5%	-7.5%	-9.5%	-0.8%	-0.8%
Percentage change inCO2 emissions v. 1990 levels (UNFCCC baseline year)[b]	+24.2%	+27.7%	+24.2%	+21.7%	+32.2%	+32.2%
Penalties for noncompliance	Same as Clean Air Act, title IV except that the excess emission penalty is three times the average market price for allowances.	Excess emission penalty equal to three times the market price for allowances on the last day of the year at issue.	To be determined by EPA.	To be determined by EPA.	$100 per excess ton plus one-for-one offset from future emissions allocations.	$100 per excess ton plus one-for-one offset from future emissions allocations.

a. S. 150 would further limit the number of emission allowances in a given year by the number of tons emitted two years prior by small electricity generating facilities, and by any number required to protect the public health, welfare, or the environment.

b. CRS calculations based on projections contained in the UNFCCC Secretariat's *2002 Climate Action Report*. Available at [http://yosemite.epa.gov/oar/globalwarming.nsf/ content/ResourceCenterPublicationsUSClimateActionReport.html]. For more information, see CRS Report RL32755, *Air Quality: Multi-Pollutant Legislation in the 109th Congress*, by Larry Parker and John Blodgett.

APPENDIX 1. CLIMATE CHANGE BILLS IN THE 109TH CONGRESS

Bill(s) and Short Title(s)	Climate Change Research	Technology Deployment	GHG Reporting and Registry	Multi-Pollutant Bill	Emissions Caps and Allowance Trading for all GHGs
ENACTED LAW					
H.R. 6, P.L. 109-58 *Energy Policy Act of 2005*		X			
SENATE BILLS					
S. 150 (Jeffords) *The Clean Power Act of 2005*			X	X	
S. 245 (Collins) *Abrupt Climate Change Research Act of 2005*	X				
S. 730 (Leahy) *Mercury Emission Act of 2005*			X	X	
S. 745 (Byrd) *International Clean Energy Deployment and Global Energy Markets Investment Act of 2005*		X			
S. 883 (Hagel) *Climate Change Technology Deployment in Developing Countries Act of 2005*		X			
S. 887 (Hagel) *Climate Change Technology Deployment and Infrastructure Credit Act of 2005*		X			
S. 1151 (McCain) *Climate Stewardship and Innovation Act of 2005*	X	X	X		X
S. 1203 (Hagel) *Climate Change Technology Tax Incentives Act of 2005*		X			

S. 2724 (Carper) *Clean Air Planning Act of 2006*			X	X	
S. 3698 (Jeffords) *Global Warming Pollution Reduction Act*	X	X	X		X
S. 342* (McCain) *Climate Stewardship Act of 2005*	X		X		X
S. 386* (Hagel) *Climate Change Technology Deployment in Developing Countries Act of 2005*		X			
S. 387* (Hagel) *Climate Change Technology Tax Incentives Act of 2005*		X			
S. 388* (Hagel) *Climate Change Technology Deployment and Infrastructure Credit Act of 2005*		X	X		
* Superseded by newer version					
HOUSE BILLS					
H.R. 759 (Gilchrest) *Climate Stewardship Act of 2005*	X		X		X
H.R. 955 (Olver) *National Greenhouse Gas Emissions Inventory Act of 2005*			X		
H.R. 1451 (Waxman) *Clean Smokestacks Act of 2005*			X	X	
H.R. 1873 (Bass) *Clean Air Planning Act of 2005*			X	X	
H.R. 2828 (Inslee) *New Apollo Energy Act of 2005*	X	X	X		X
H.R. 5049 (Udall, T.) *Keep America Competitive Global Warming Policy Act of 2006*	X		X		X
H.R. 5642 (Waxman) *Safe Climate Act of 2006*			X		X

Appendix 2. Key Provisions of Climate Change Legislation in the 109th Congress

Bill No.	Sponsor	Major Actions	Key Provisions
H.R. 6, P.L. 109-58	Barton	Introduced April 18, 2005; passed House April 21, 2005; passed Senate June 28, 2005; conference report file July 27, 2005; agreed to in House July 28; agreed to in Senate July 29; signed into law August 8, 2005.	Omnibus energy bill addressing various climate- and non-climate related topics. Among other provisions, establishes loans, loan guarantees, etc. to deploy technology for greenhouse gas intensity reduction (similar language to S. 887); requires the Secretary of State to provide assistance to developing countries on projects to reduce greenhouse gas intensity; establishes an export initiative for greenhouse gas reduction technology (similar language to S. 883)
SENATE BILLS			
S. 150	Jeffords	Introduced January 25, 2005; referred to Senate Environment and Public Works.	Amends the Clean Air Act to require the Administrator of the Environmental Protection Agency to promulgate regulations to achieve specified reductions in emissions of sulfur dioxide, nitrogen oxides, carbon dioxide and mercury from certain electric generation facilities by January 1, 2010.
S. 245	Collins	Introduced February 1, 2005; referred to Senate Commerce, Science, and Transportation.	Establishes within the Department of Commerce a research program on abrupt climate change.
S. 730	Leahy	Introduced April 6, 2005; referred to Senate Environment and Public Works.	Amends the Clean Air Act to require the Administrator of the Environmental Protection Agency to promulgate regulations to achieve specified reductions in emissions of sulfur dioxide, nitrogen oxides, carbon dioxide and mercury from certain electric generation facilities by January 1, 2010.
S. 745	Byrd	Introduced April 11, 2005; referred to Senate Foreign Relations.	Establishes within the Department of State a program to assist developing countries in the demonstration and deployment of emission reduction technologies.
S. 883	Hagel	Introduced April 21, 2005; referred to Senate Foreign Relations — see also H.R. 6 (Senate Version)	Requires the Secretary of State to provide assistance to developing countries on projects to reduce greenhouse gas intensity; establishes an export initiative for greenhouse gas reduction technology.

S. 887	Hagel	Introduced April 21, 2005; referred to Senate Energy and Natural Resources — see also H.R. 6 (Senate Version)	Establishes loans, loan guarantees, etc. to deploy technology for greenhouse gas intensity reduction.
S. 1151	McCain	Introduced May 25, 2005; referred to Senate Environment and Public Works.	Requires any entity that emits more than 10,000 metric tons of greenhouse gases (CO_2 equivalent) to reduce emissions to year 2000 levels by 2010. Allows: tradeable credits for reductions beyond those required, reductions from non- covered entities, increases in carbon sequestration, and emissions reductions in other countries. Promotes innovation on mitigation technologies and establishes incentives for technology deployment.
S. 1203 S. 2724	Hagel Carper	Introduced June 8, 2005; referred to Senate Finance. Introduced May 4, 2006; referred to Senate Environment and Public Works.	Establishes tax credits for investment in technologies to reduce greenhouse gas intensity; also provides tax incentives for nuclear technologies. Amends the Clean Air Act to require the Administrator of the Environmental Protection Agency to promulgate regulations to achieve specified reductions in emissions carbon dioxide and pollutants from certain electric generation facilities by 2010 (2007 for nitrogen oxides).
S. 3698	Jeffords	Introduced July 20, 2006; referred to Senate Environment and Public Works.	Amends the Clean Air Act to require the Administrator of the Environmental Protection Agency to promulgate regulations to achieve an 80% reduction in greenhouse gas emissions below 1990 levels by 2050; establishes efficiency and/or emissions standards for various sectors; promotes research and development
S. 342*	McCain	Introduced February 10, 2005; referred to Senate Environment and Public Works.	Requires any entity that emits more than 10,000 metric tons of greenhouse gases (CO_2 equivalent) to reduce emissions to year 2000 levels by 2010. Allows: tradeable credits for reductions beyond those required, reductions from non- covered entities, increases in carbon sequestration, and emissions reductions in other countries.
S. 386*	Hagel	Introduced February 15, 2005; referred to Senate Foreign Relations.	Requires the Secretary of State to provide assistance to developing countries on projects to reduce greenhouse gas intensity; establishes an export initiative for greenhouse gas reduction technology.

APPENDIX 2. (CONTINUED)

Bill No.	Sponsor	Major Actions	Key Provisions
S. 387*	Hagel	Introduced February 15, 2005; referred to Senate Finance.	Establishes tax credits for investment in technologies to reduce greenhouse gas intensity; also provides tax incentives for clean coal and nuclear technologies.
S. 388*	Hagel	Introduced February 15, 2005; referred to Senate Energy and Natural Resources.	Establishes a loans, loan guarantees, etc. to deploy technology for greenhouse gas intensity reduction; establishes a voluntary national greenhouse gas registry.
*Superseded by newer version			
HOUSE BILLS			
H.R. 759	Gilchrest	Introduced February 10, 2005; referred to House Science, and House Energy and Commerce.	Requires any entity that emits more than 10,000 metric tons of greenhouse gases (CO_2 equivalent) to reduce emissions to year 2000 levels by 2010. Allows: tradeable credits for reductions beyond those required, reductions from non- covered entities, increases in carbon sequestration, and emissions reductions in other countries.
H.R. 955	Olver	Introduced February 17, 2005; referred to House Energy and Commerce.	Requires EPA to establish a GHG emissions information system to collect information submitted regarding an entity's GHG emissions. Establishes mandatory registry for entities that emit more than 10,000 metric tons of carbon dioxide equivalent.
H.R. 1451	Waxman	Introduced March 17, 2005; referred to House Energy and Commerce.	Amends the Clean Air Act to require the Administrator of the Environmental Protection Agency to promulgate regulations to achieve specified reductions in emissions of carbon dioxide and pollutants from certain electric generation facilities by 2010.
H.R. 1873	Bass	Introduced April 27, 2005; referred to House Energy and Commerce.	Amends the Clean Air Act to require the Administrator of the Environmental Protection Agency to promulgate regulations to achieve specified reductions in emissions carbon dioxide and pollutants from certain electric generation facilities by 2010 (2009 for nitrogen oxides).
H.R. 2828	Inslee	Introduced June 8, 2005; referred to House	Omnibus energy bill addressing various climate- and non-climate

		Energy and Commerce, among other committees.	related topics. Among other provisions, requires any entity that emits more than 10,000 metric tons of greenhouse gases (CO_2 equivalent) to reduce emissions to year 2000 levels by 2010. Allows: tradeable credits for reductions beyond those required, reductions from non-covered entities, increases in carbon sequestration, and emissions reductions in other countries.
H.R. 5049	Udall, T.	Introduced May 24, 2006; referred to House Energy and Commerce, among other committees.	Establishes a system of tradeable allowances for greenhouse gas emissions from fossil fuel supply and combustion, with a maximum “safety valve” price of $25 per ton of carbon, adjusted for inflation. Establishes an Advanced Research Projects Agency within the Department of Energy.
H.R. 5642	Waxman	Introduced June 20, 2006; House Energy and Commerce, among other committees.	Amends the Clean Air Act to require the Administrator of the Environmental Protection Agency to promulgate regulations to achieve an 80% reduction in greenhouse gas emissions below 1990 levels by 2050.

ENDNOTES

[1] Under the United Nations Framework Convention on Climate Change (UNFCCC), greenhouse gases include carbon dioxide (CO_2, the most ubiquitous and primary greenhouse gas), methane (CH_4), nitrous oxide (N_2O), hydrofluorocarbons (HFCs), perfluorocarbons (PFCs), and sulfur hexafluorane (SF_6). Some other greenhouse gases are controlled under the Montreal Protocol on Substances That Deplete the Ozone Layer.

[2] Greenhouse gas intensity is a measure of the amount of carbon dioxide (or equivalent) emitted per unit of gross domestic product.

[3] For more information on the science and policy of Global Climate Change see CRS Issue Brief IB89005, *Global Climate Change*, by John R. Justus and Susan R. Fletcher.

[4] This report does not include bills with other focuses that also had research components related to climate change (particularly sequestration, renewable energy, and energy efficiency).

[5] For more information, see CRS Report 98-235, *Global Climate Change: U.S. Greenhouse Gas Emissions — Status, Trends, and Projections*, by John Blodgett and Larry Parker.

[6] U.S. Environmental Protection Agency, *U.S. GHG Emissions and Sinks 1990-2001*, p. ES-6. Additional sources are agriculture (7.6%), commerce (7.2%), and residential activities (5.4%).

[7] For more information, see Pew Center on Global Climate Change, *Climate Change Activities in the United States: 2004 Update*, Arlington, VA, 2004.

[8] Defined as any product for which the Department of Energy has promulgated final regulations for energy efficiency, energy conservation, maximum energy use, or energy consumption.

[9] The acid rain program caps emissions from each source, but allows sources to exceed their caps if they purchase credits from sources that achieve emissions reductions beyond those required.

[10] S. 131 (Inhofe) and H.R. 227 (Sweeny) would also establish a cap-and-trade program for nitrogen oxides, sulfur dioxide, and mercury from utilities. However, the bills do not address carbon dioxide emissions.

[11] For more information on multi-pollutant bills, see CRS Report RL32755, *Air Quality: Multi-Pollutant Legislation in the 109th Congress*, by Larry Parker and John Blodgett.

In: Global Climate Change Revisited
Editor: Horace M. Karling, pp. 147-153
ISBN 978-1-59454-039-4

Chapter 9

GLOBAL CLIMATE CHANGE AND WILDLIFE*

Pervaze A. Sheikh, M. Lynne Corn, Jane A. Leggett and Peter Folger

SUMMARY

Recently projected climate changes could have widespread effects on wildlife species. These effects may include extinctions, changes in species' ranges, mismatches in their phenology (timing of pollination, flowering, etc.), and population declines. Other species could flourish in response to projected climate change. Effects of climate change on wildlife have been reported for some species and populations in localized areas. Many studies also show that climate change acts in concert with other variables to effect changes in species. If climate change is widespread, there is uncertainty on how wildlife will adapt. Some suggest that evolution and migration will enable species to adapt, whereas others contend that adaptation will be minimal because of limited habitat and that changes in climate may occur more rapidly than adaptation can respond to.

INTRODUCTION

Climate change can have major repercussions for ecological systems, as indicated by past major climate shifts. Effects of climate change are sometimes local, although some may be widespread and linked to changes in other regions through food chains, nutrient flows, and atmospheric and ocean circulation, among other things. Some populations and species could flourish as the environment changes, whereas others could decline. Some species would probably be able to adapt or adjust their ranges to stay within a hospitable habitat. Others may be affected by obstacles to migration, disruptions in food chains, or habitat alteration or loss.

Wildlife species can be affected by several climatic variables such as increasing temperatures, changes in precipitation, and extreme weather events. Some scientists predict

* Excerpted from CRS Report RS22597 dated February 16, 2007

that climate change can cause extinctions, which would lower biodiversity. Other scientists contend that many species could benefit from climate change, such as some migratory bird species which may have shorter distances to migrate for breeding and over-wintering.[1]

The impacts of regional climate change and extreme weather on wild species has been studied for several decades.[2] Paleoclimatic studies have shown that species have adjusted to climate changes at times in the past without mass extinctions.[3] Yet, it is uncertain if projected climate change widely foreseen today would mimic climate change events in the geological record, and if human-ecosystem relationships would help or hurt adaptation. In the last ten years, scientists have documented the effects of climate change on species and populations on every continent and in most taxonomic groups.[4] Some studies provide correlations between climate change and species changes, others predict changes with climate and species models, and some demonstrate mechanistic connections between species changes and climate change.[5] Many studies report that climate change acts in concert with other factors to affect species and their habitat. For this reason, and because of differences across biomes and species, it is difficult to generalize about the overall impacts of climate change on biodiversity, wildlife, and ecosystems.

In discussing the effects of climate change on wildlife, some contend that climate change may not be bad for species, even if some go extinct. They reason that because biodiversity is greatest in tropical zones, biodiversity may benefit by warming and greater precipitation. Further, they argue that climate change may induce species to evolve behaviors or traits that will allow them to adapt to different conditions.[6] Others counter these arguments by stating that climate change may be too rapid for species to evolve. Scientific studies show that climate change affects species in different ways, yet the message that these studies convey may be biased because studies that show no changes in species due to climate change are less likely to be published.[7] With monitoring, research, and modeling, the responses of populations, species, landscapes and ecosystems to climate change may be predicted more reliably, but the complexity of living systems and the lack of long-term data make this task difficult.

Climate Change Affecting Wildlife

Climate change producing higher temperatures may affect wildlife by causing species to expand, contract, or shift their ranges, typically toward the poles or higher elevations.[8] It may also alter their phenology[9] leading to mismatches with their food and habitat resources, cause habitat loss, and allow exposure to new pathogens. These changes could have either negative or positive effects on the abilities of species to survive. The abilities of species to adapt to some degree of climate change, in part, depends upon the rate at which the change occurs. The following sections of this report provides examples of wildlife that are thought to be, in part, affected by climate change. Examples were taken from published studies with appropriate citations. The examples illustrate the types of changes that have been observed.

Range Shifts

There are few studies that report the effects of climate change over an entire species' range, partly because of the difficulties of gathering data throughout its range and for a period sufficient to produce significant results.[10] Even so, range shifts have been documented in some species of birds and insects, and in some marine communities. One example is the mountain pine beetle (*Dendroctonus ponderosae*). The mountain pine beetle feeds primarily on lodgepole pine,[11] and occasionally on other species including whitebark pine (*Pinus albicaulis)*.[12] Until recent decades, the beetle has been held in check at higher elevations, due to its inability to complete a full life cycle during the short growing season. Research shows that, with warming, the beetles are becoming able to complete a full reproductive cycle in whitebark pine habitat.[13] As a result, they can time their breeding cycle to produce huge numbers of eggs whose emerging larvae can overwhelm the defenses not only of damaged trees but also of healthy ones. The beetles' expansion correlates closely with local warming data. This beetle is reported to be moving up mountains, and also northward into Canada where it is occupying some forest stands for the first time in recorded history.

Phenological Changes

Phenological cues are critical in the life cycle of many species. When global warming alters the climate, certain phenological events generally shift accordingly. If a phenological change in one species does not match linked changes in an interdependent species, an *ecological mismatch* can occur.[14] For example, if certain trees bloom earlier in response to warmer springs, but pollinators do not hatch earlier, disruptions (e.g., failed or inadequate pollination) may occur. Mismatches can occur between predators and prey, herbivorous insects and host plants, and pollinators and flowering plants, among others. Scientists are studying potential mismatches caused by climate change because they can disrupt the reproductive cycle of many species. In one example, variability in precipitation, linked to regional climate change, is considered as the cause of the extinction of two populations of checkerspot butterflies in California.[15] Changing levels of precipitation were found to alter the relationship between butterfly larvae and host plants. In very wet or dry years, larvae do not get the opportunity to feed on host plants before the plants die. These lost opportunities for successful reproduction cause variation in butterfly abundance, which eventually may lead to their extinction.

Population Decline due to Habitat Loss

There are few examples of studies that directly associate wildlife population declines with climate change. Indirect links have been reported, such as population declines in species due to habitat loss exacerbated by climate change. For example, the impact of climate change on Antarctic krill (small, shrimp-like animals) has been documented. Antarctic krill are major grazers in the Southern Atlantic Ocean and are food for various species of fish targeted by commercial fisheries. Declines in sea-ice area and in the duration of sea-ice formation, linked in part to climate change in Antarctica, have been connected to declines in krill density.

Variability in krill densities has implications for the Antarctic food web, which also includes penguins, albatrosses, seals, and whales.[16]

Pathogens

Some pathogens could increase their ranges due to climate change, whereas others may diminish their ranges. For example, harlequin frogs are declining due to increased outbreaks of a harmful chytrid fungus. In the last 20 years, 67% of the 110 species of harlequin frogs have gone extinct.[17] A mechanistic explanation of how climate change promotes outbreaks of the chytrid fungus has been presented. Nighttime temperatures are shifting closer to the thermal optimum for chytrid growth, and daytime cloudiness prevents frogs from seeking thermal refuges.[18] These climatic shifts are promoting outbreaks of chytrid fungus in areas where harlequin frog populations reside.

Wildlife Adaptation to Climate Change

Species and populations have demonstrated that they can adapt in a variety of ways to some degree of climate change. For example, they can shift their ranges, change their phenologies, and with sufficient time, may adapt to climate change through evolution. Paleoclimate evidence indicates that some ecosystems restructured during past climate change without records of mass extinctions.[19] Assessing the ability of species to adapt to climate change is difficult because other stressors that may drive adaptation could be present, such as habitat loss due to factors other than climate change, competition from other species, and invasive species. Further signs of adaptation, such as range expansions or shifts in phenology, may initially be viewed as positive responses by a species to climate changes, but have negative consequences if these shifts do not correspond to other ecosystem properties or other necessary conditions needed for that species' survival. For example, if the natural response to climate change for a species would be to expand northward, the consequences might be severe if intervening developed areas replace suitable habitat or block access to it.

The rate of climate change reported in this century is generally seen as being uncharacteristically rapid and many scientists have questions concerning how or whether species will adapt. Climate change models based on global average temperature changes are less precise about local climate change that could be important for many species. This uncertainty has led some to propose artificial methods of adaptation such as *assisted migration*, which is the collection and transport of species from unsuitable to suitable habitat. This technique has been used with mixed results with endangered species and many question its viability and cost. Other suggestions to assist adaptation of wildlife include building corridors between protected areas to give species paths to migrate and shift their ranges.

Policy Implications

Climate change has the potential to adversely affect species. Some contend that effects from recently projected climate change will warrant mitigation of effects to wildlife and that planning for mitigation is prudent now. Others assert that the effects of climate change on species may not need intervention and that many species may be able to adapt successfully. Addressing this issue may require answering several questions that are still being discussed. For example, will the rate of any climate change be slow enough to allow species to adapt? What are local changes in climate that would affect specific populations and ecosystems? How could human development and management of natural resources interact with the natural adaptation process of species under climate change? And, is the extinction of species due to climate change a concern or a natural process?

If the effects of climate change on wildlife are widespread and occur rapidly, there are resulting legislative implications. Several wildlife laws and management plans that have been developed with an understanding of the life history, distribution, and habitat needs of species may merit assessment to see if they function as desired under a projected climate change. For example, under the Endangered Species Act (ESA), critical habitat designation, habitat conservation plans, and biological opinions for several species might have to be re-done if data change. This could overwhelm scientific resources and the resources of the pertinent agencies to review new data. For aquatic species, changes in their habitat requirements could require water flows and water supply to be altered in river management plans. This possibility is tempered by some who contend that changes to species may be gradual enough that wildlife laws that address potential changes may be sufficient. In the case of ESA, there are provisions that require the revision of recovery plans or conservation plans if new scientific data are presented.

The control, eradication, and prevention of invasive and pest species are generally addressed on a species-by-species basis in management plans. If rapid climate change affects pest or exotic species, scientific knowledge may not keep pace. Further, changes in species ranges may make it difficult to determine if they are exotic species or native species with new ranges. Separating exotic species from adapting species may be relevant for national park and refuge managers. Managers might be faced with challenging management objectives: whether to preserve and protect current wildlife species and facilitate the adaptation of new species into protected areas.[20]

The effects of climate change on wildlife are being viewed by some as an issue within the larger context of potential changes to ecosystems. Ecological factors that affect wildlife, including water flows and availability, soils, and nutrients, can be addressed in an ecosystem context. Consequently, large-scale ecosystem restoration initiatives might have to incorporate changes in their restoration plans to address the potential effects of climate change. Restoration plan horizons generally range from 20 to 50 years over which local climate may change substantially. For example, the restoration of the Everglades ecosystem is expected to take 30 years under the Comprehensive Everglades Restoration Plan. This plan could need to be revised if recently predicted sea level rise occurs and causes changes in water flow in the Everglades.[21]

ENDNOTES

[1] R.A. Robinson, et al. *Climate Change and Migratory Species*, British Trust for Ornithology (Norfolk, UK: 2005), 304 p.

[2] Camille Parmesan, "Ecological and Evolutionary Responses to Recent Climate Change," *Annual Review of Ecology, Evolution, and Systematics*, vol. 37 (2006): 637-669. Hereafter referred to as *Parmesan 2006.*

[3] Isabel P. Montanez, et al., "CO*2*-Forced Climate and Vegetation Instability During Late Paleozoic Deglaciation," *Science*, vol. 315, no. 5808 (2007): 87-91.

[4] *Parmesan 2006.*

[5] T. Root, et al., Fingerprints of Global Warming on Wild Animals and Plants, *Nature*, vol. 421(2003): 57-60.

[6] A.C. Baker et al., Corals Adaptive Response to Climate Change, *Nature*, vol. 430 (2004): p. 741.

[7] See *Parmesan 2006* for a discussion on this issue.

[8] L. Hannah, T.E. Lovejoy, and S.H. Schneider, "Biodiversity and Climate Change in Context," in *Climate Change and Biodiversity*, T.E. Lovejoy and L. Hannah, eds. (Yale University Press, New Haven CT: 2005): 3-14.

[9] *Phenology* is "the relationship between climate and periodic natural phenomena such as the migration of birds, bud bursting, or the flowering of plants." *The Dictionary of Ecology and Environmental Science*, H. W. Hart, ed. (New York, NY: Henry Holt & Co, 1993)

[10] *Parmesan 2006.*

[11] The lodgepole pine is adapted to this beetle, and stands of this fast-growing tree can regenerate fairly quickly — for trees.

[12] The heavy seed crop of the whitebark pine supports many species, including the Clark's nutcracker which harvests and stores large caches of the seeds, and feeds on them through winter and into the breeding season. Red squirrels also collect and store the seeds. The stored caches are important fall and spring food sources for grizzlies.

[13] This discussion is adapted from Michel Nijhuis, "Global Warming's Unlikely Harbingers," *High Country News*, vol 36 (July 19, 2004): p. 1. The author cites a variety of primary sources, largely U.S. Forest Service scientists.

[14] Marcel Visser and Christiaan Both, "Shifts in Phenology Due to Global Climate Change: The Need for a Yardstick," *Proceedings of the Royal Society Bulletin,* vol. 272 (2005): 2561-2569.

[15] John McLaughlin, et al., "Climate Change Hastens Population Extinctions," *Proceedings of the National Academy of Sciences,* vol. 99, no. 9 (2002): 6070-6074.

[16] Angus Atkinson, et al., "Long-term decline in krill stock and increase in salps within the Southern Ocean, *Nature*," vol. 432 (2004): 100-103.

[17] J. Alan Pounds, et al., "Widespread Amphibian Extinctions from Epidemic Disease Driven by Global Warming," *Nature,* vol. 439 (2006): 161-167.

[18] Andrew Blaustein and Andy Dobson, "A Message from the Frogs," *Nature* vol. 439 (2006): 143-144.

[19] T.E. Lovejoy and L. Hannah, "Global Greenhouse Gas Levels and the Future of Biodiversity," in *Climate Change and Biodiversity*, T.E. Lovejoy and L. Hannah eds. (Yale University Press, New Haven CT: 2005): 387-395.

[20] D. Scott, "Integrating Climate Change into Canada's National Park System," in *Climate Change and Biodiversity*, T.E. Lovejoy and L. Hannah eds. (Yale University Press, New Haven CT: 2005): 342-345.

[21] Some restoration plans include adaptive management, which incorporates information from new or unforeseen circumstances into a project to assure that the goals of the project are achieved efficiently. It is unclear if adaptive management will allow for large-scale changes in plans that may happen under climate change.

INDEX

A

B

C

D

E

F

G

H

I

Q

R

S